A Guide to Computing Statistics with SPSS for Windows

A Guide to Computing Statistics with SPSS for Windows

DENNIS HOWITT AND DUNCAN CRAMER

PRENTICE HALL
HARVESTER WHEATSHEAF
LONDON NEW YORK TORONTO SYDNEY TOKYO
SINGAPORE MADRID MEXICO CITY MUNICH PARIS

First published 1997 by
Prentice Hall/Harvester Wheatsheaf
Campus 400, Maylands Avenue
Hemel Hempstead
Hertfordshire, HP2 7EZ
A division of
Simon & Schuster International Group

Typeset in 10/12pt Times
by Dorwyn Ltd, Rowlands Castle, Hants

Printed and bound in Great Britain by
Redwood Books, Trowbridge, Wiltshire

For information about SPSS contact: SPSS UK Ltd., 1st Floor, St Andrew's House, West St., Woking, Surrey GU21 1EB Tel: 01483 719200 Fax: 01483 719291

Library of Congress Cataloging-in-Publication Data

Howitt, Dennis.
A guide to computing statistics with SPSS for Windows / Dennis Howitt and Duncan Cramer.
p. cm.
ISBN 0–13–729197–3 (pbk. : alk. paper)
1. SPSS for Windows. 2. Sociology–Statistical methods–Computer programs. I. Cramer, Duncan, 1948– . II. Title.
HM48.H68 1996
300'.285'5369–dc20 96–42185
CIP

British Library Cataloguing in Publication Data

A catalogue record for this book is available from the British Library

ISBN 0–13–729197–3 (pbk)

2 3 4 5 01 00 99 98

Contents

Introduction

This book can be used as a stand-alone practical guide to SPSS for Windows when analysing psychological and similar data. However, it draws on the examples of research described and explained in *An Introduction to Statistics in Psychology* (Howitt and Cramer, 1997), an introductory but extended textbook on psychological statistics. In the present volume this is referred to as *ISP*. Some readers will need to supplement the present guide by consulting the textbook. Others, especially those already familiar with psychological statistics, may find this guide sufficient most of the time.

SPSS (Statistical Package for the Social Sciences) was initially developed in 1965 at Stanford University in California. *Windows* is a system used on personal computers which is operated by selecting options from *menus* and *dialog boxes*. In the interest of simplicity, we have confined our coverage to menus and dialog boxes alone. You can, however, also enter typed commands or instructions. The big advantage of menus is that you can see the options on the screen and do not need to remember them or refer to a manual.

This guide is based on the latest version of SPSS for Windows (i.e. Release 6 for Windows Version 3.1 and Release 7 for Windows 95; although these are not identical, they are so similar that there should be little problem in using either version). Usually the data and statistical analyses carried out correspond to those used in *An Introduction to Statistics in Psychology*. Consult that book for more detailed explanations of these statistical applications.

Reference

Howitt, D., and Cramer, D. (1997). *An Introduction to Statistics in Psychology: A Complete Guide for Students*. Hemel Hempstead: Prentice Hall/Harvester Wheatsheaf.

Chapter 1

How to access SPSS for Windows and enter data

This chapter gives an overview of the basic operation of SPSS for Windows. It includes data entry as well as saving data as files.

1.1 Introduction

SPSS for Windows is accessed on a personal computer (or *PC* for short). A PC consists of five major components (Figure 1.1):

1. A television-like *screen* (also called a *monitor* or *visual display unit*) to display information.
2. A *keyboard* to type in information.
3. A *system unit* which houses the computer itself and usually a *drive* for inserting a *floppy disk*.

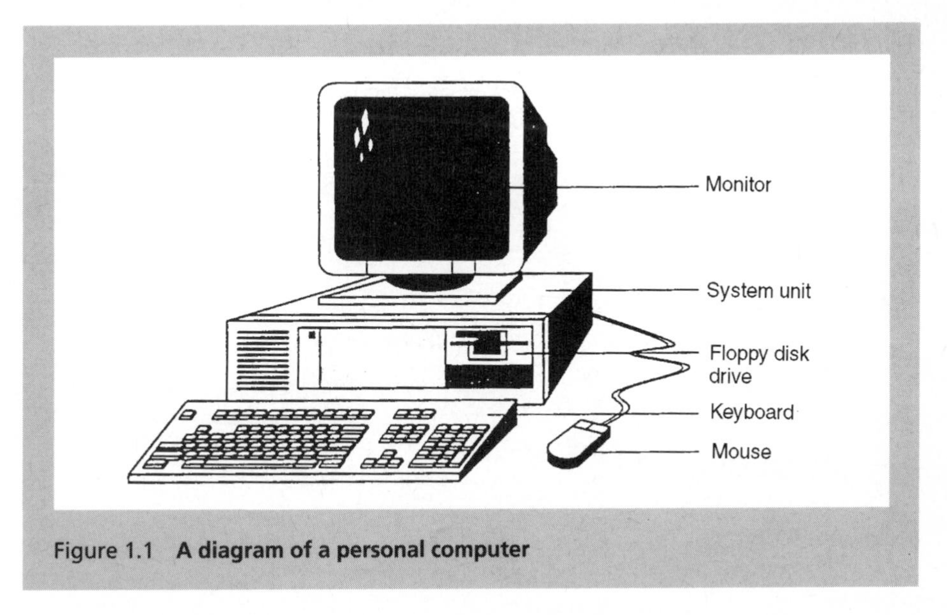

Figure 1.1 **A diagram of a personal computer**

4. A *printer* to print information you have produced.
5. Usually a small hand-held 'switch' called a *mouse*.

Moving the mouse on a hard surface causes the *pointer* to move on the screen. This pointer is usually called a *cursor*. Selecting your chosen option involves moving the cursor to cover that option on the screen. You then press or *click on* a particular option using the *left* button of the mouse.

At a university, college or school, you will normally need:

1. A personal code (or *ID*).
2. A *password*.

Obtain these from the appropriate individual or department (e.g. the computer centre) at your institution; the ID and password have to be typed in before you can access SPSS.

1.2 To access SPSS

- Before switching on the computer, make sure you remove floppy disks left in any of the disk drives. If you forget, you may have to switch off and start again.
- Make sure that both the screen *and* the system unit are switched on. Usually this involves just one switch.
- What to do next depends on the way your computer has been set up, so precise instructions cannot be given here. They may involve more than one step; usually you have to enter your *ID* and *password*.

> To remind yourself, write here the additional steps you need to take to access SPSS at your particular location:
>
> - ____________________
> - ____________________
> - ____________________
> - ____________________
>
> Remember to keep your password secret!

A *Program Manager Window* of more or less the kind shown in Figure 1.2 should appear once you have completed the necessary steps. (The Window may not be identical to the one in Figure 1.2. It depends on local circumstances. Be careful to look for the Program Manager bar at the top of the screen. The rest of

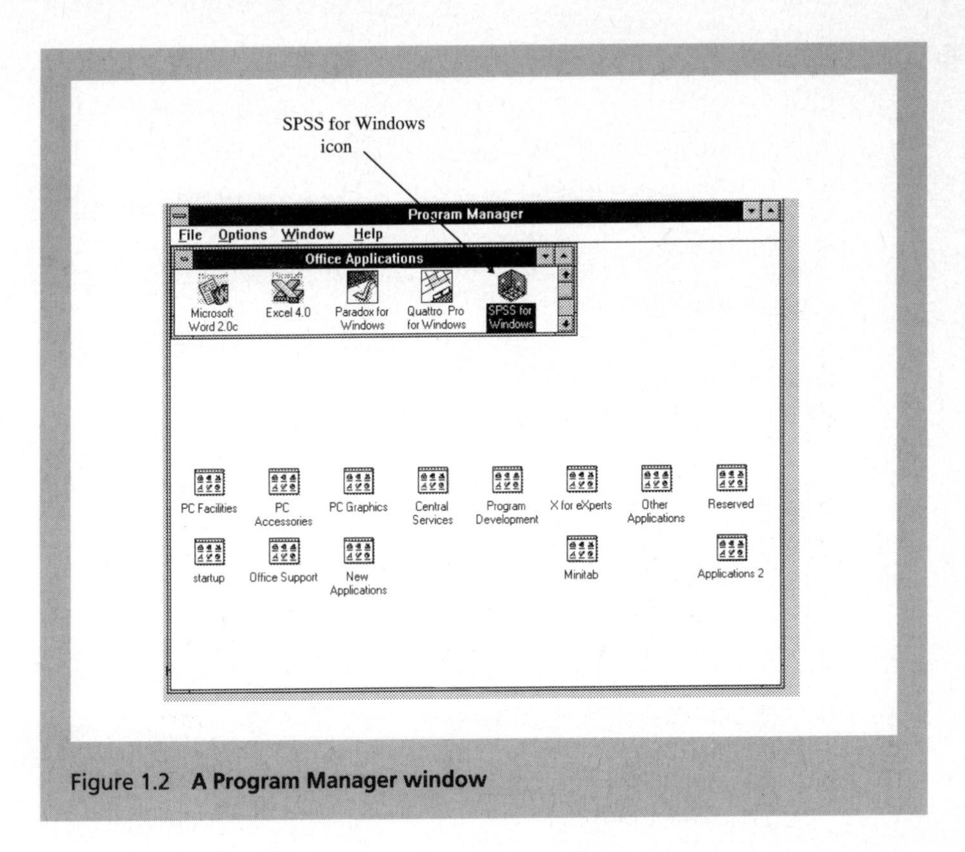

Figure 1.2 **A Program Manager window**

the screen may vary somewhat from our example depending on how it has been set up or *configured*.)

- Move the cursor to the SPSS for Windows *icon* or symbol (Figure 1.2) on the screen.
- *Quickly* double-press or double-click the left button on the mouse.

If nothing changes on the screen after a few seconds, quickly double-click the left-hand mouse button again. Repeat if necessary – some people find double-clicking awkward at first.

Selecting the SPSS icon produces, after some seconds' delay, the screen depicted in Figure 1.3. This consists of three overlapping windows identified as follows on the top bar of each window:

1. 'SPSS for Windows' (the Applications window).
2. '!Output1'.
3. 'Newdata'.

The form of the cursor will vary according to which window it is in. It may be one of the following.

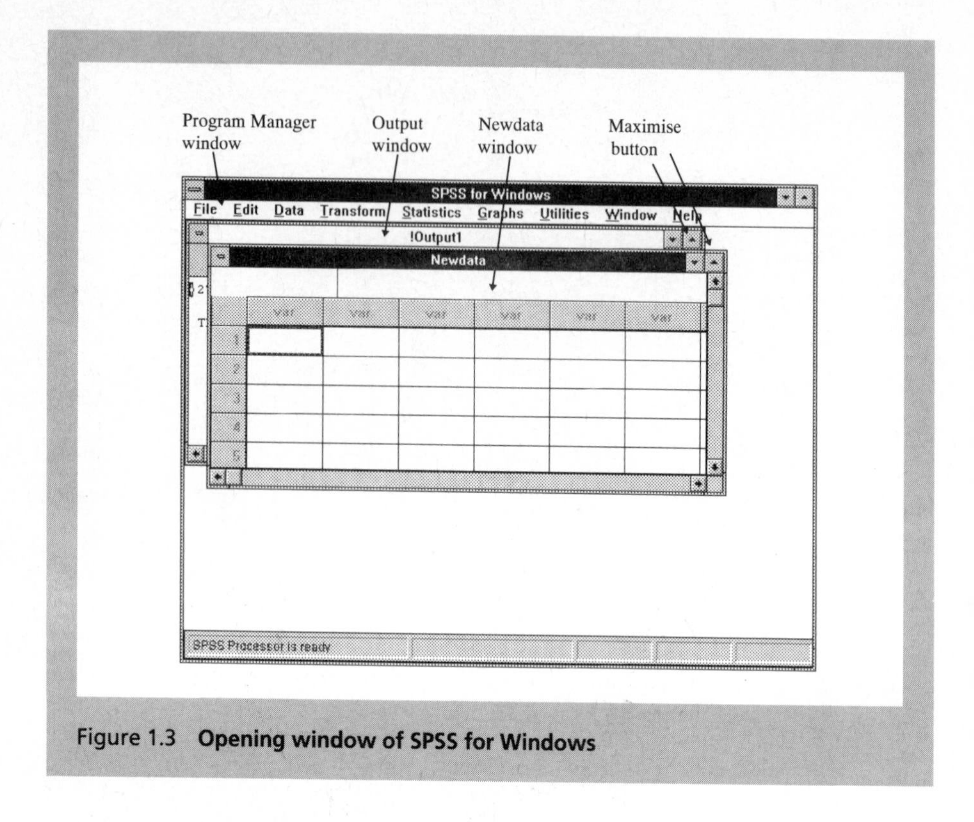

Figure 1.3 **Opening window of SPSS for Windows**

1. A pointer in the Application window.
2. A thin upright cross (sometimes called a cross-hair or wire) in the Newdata window.
3. A flashing vertical line in the Output window.

1.3 To enter data

Data are put in the Newdata window. We will enter the ages of 12 university students shown in Table 1.1 (*ISP* Table 3.7).

Quantitative data are normally organised in terms of the columns and rows of a table:

1. Different columns represent different variables such as age and sex in SPSS.
2. Different rows represent the data for different cases. In psychology a case is usually a person.

Because age is our only variable, only one column is needed for the 12 values. We will use the first column to contain the age scores but any column could be used.

Table 1.1 **Twelve numerical scores**

18
21
23
18
19
19
19
33
18
19
19
20

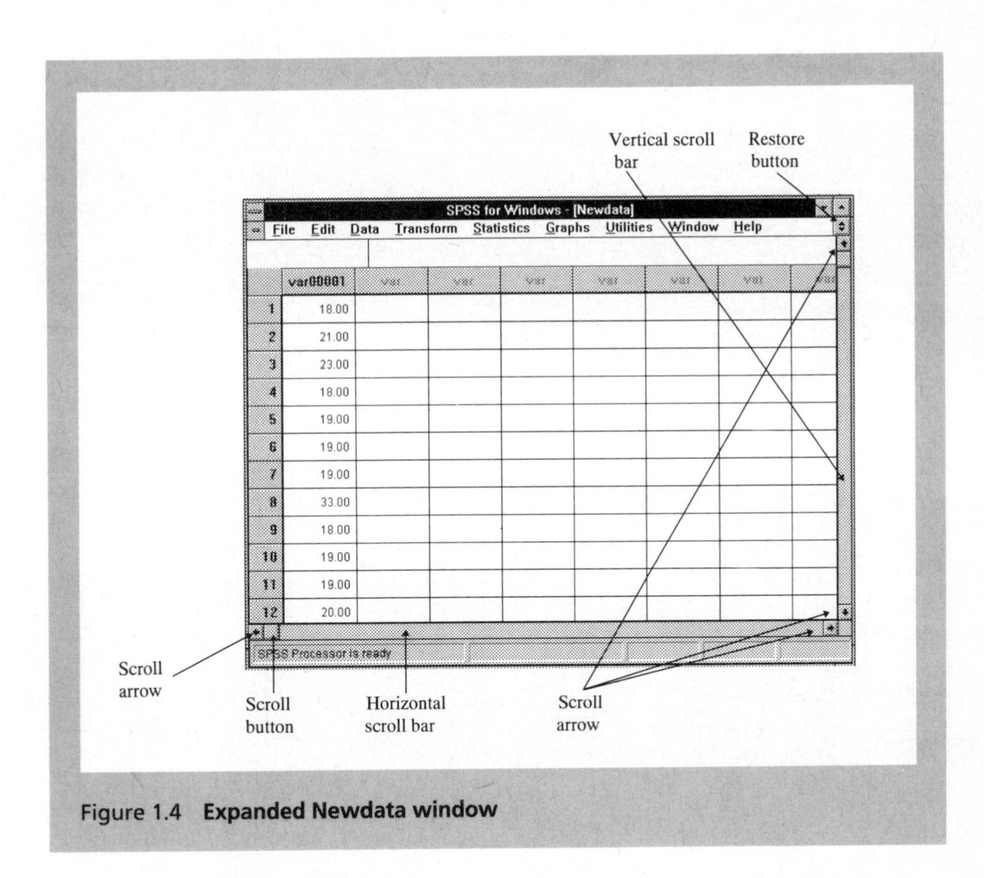

Figure 1.4 **Expanded Newdata window**

The *cell* in the first row of the first column is *framed in bold* (Figure 1.3). The bold frame denotes where the datum will go once it has been typed in on the keyboard and the *Return* or *Enter* key pressed.

- Type in the first number, 18, on the keyboard.
- Press Return or Enter. 18 will be placed in the first cell after a short delay. Wait until the number appears in the cell before typing in the next number.
- The name of the column will change from 'var' to 'var00001' and the bold frame of the cell will move to the next cell below (i.e. the cell in the second row of the first column).
- This new cell is now active. Type in all 12 values (Figure 1.4). (The window has been expanded as explained in 1.4 below.)

Note that the values are to two decimal places. SPSS does this automatically or *by default* unless you instruct it otherwise. In this case, age is a whole number so the two decimal places will remain as zeroes and have no effect.

Mistakes in Newdata

If you type in the wrong numbers then move to that cell and type in the right values.

If you forget to put in a row, such as missing out the value of 21 in the second row of Newdata, then

1. Move the cursor to that cell and press the left button.
2. Move the cursor to the Data option on the menu bar of the Applications window and press the left button which produces a drop-down menu (Figure 2.2).
3. Move the cursor to the Insert Case option on this drop-down menu and press the left button.
4. Type in the correct value (e.g. 21).
5. Press the Return key.

1.4 Changing window size

To expand any window, including Newdata:

- Move the cursor onto the Maximise 'button' ▲. This is the upward-pointing arrowhead in the upper right-hand corner of the window (Figure 1.3).
- Press the left button of the mouse.

- To revert the window to its original size move the cursor onto the lower of the two 'buttons' called the Restore button ♦ (Figure 1.4). This 'button' has an upward and downward pointing arrowhead. Press the left button of the mouse.

1.5 Moving within a window with the mouse

There may be more information in a window than is visible at any one time (as in the unexpanded Newdata window):

- To move or to *scroll* vertically use the *vertical scroll bar* (Figure 1.4).
- To scroll horizontally use the *horizontal scroll bar* (Figure 1.4).

The position of the scroll button or box (Figure 1.4) indicates your relative position in the window. For example, if the scroll box is at the top of the scroll bar (as it is in Figure 1.4), you are at the top of the window. If the scroll box is in the middle of the scroll bar (as it is in Figure 1.5) you will be in the middle of the window.

- To scroll one screenful at a time place the cursor on the *scroll arrow* pointing in the direction you want to go and press the left mouse button once. For example, to move down a screenful at a time place the cursor on the downward-pointing arrow (or below the scroll box) and press the left mouse button.

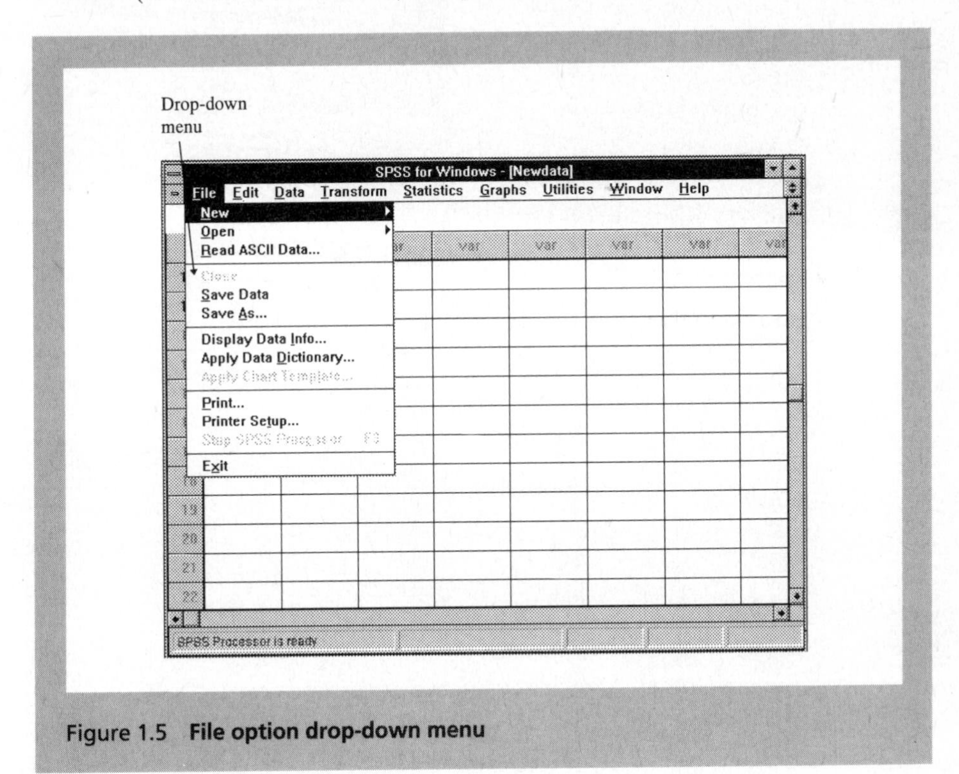

Figure 1.5 **File option drop-down menu**

- You can also *drag* the scroll box to the position in the window you want by placing the cursor on it, pressing the left mouse button without releasing it, moving the cursor to the desired position, and then releasing the button.

1.6 Moving within a window using keyboard keys

You can scroll within a window using the *cursor keys* (on the keyboard) which as their name implies affect the cursor. Although you can manage without most of them, check the following keyboard keys to see their functions:

- The left-facing arrow ← moves the cursor one space or character to the left while the right-facing arrow → moves the cursor one space or character to the right.
- The upward-facing arrow ↑ moves the cursor one line up and the downward-facing arrow ↓ one line down.
- Press the 'Home' key to move the cursor to the beginning of the line it is on and the End key to move the cursor to the end of that line.
- Press the PgUp (Page Up) key to move the cursor up one screen or page at a time.
- Press the PgDn (Page Down) key to move it down one screen at a time.
- Holding down the Ctrl (Control) key *and* pressing PgUp moves the cursor to the top left of the screen.
- Holding down the Ctrl key *and* pressing the PgDn key takes the cursor to the bottom right of the screen.
- Holding down the Ctrl key *and* pressing the Home key moves the cursor to the top left of the window.
- Holding down the Ctrl key *and* pressing the End key moves the cursor to the bottom left of the window.

1.7 To save data

Data are stored in *files*. Saving your data in files ensures that you do not waste time retyping it. Files can be stored:

1. on the *hard disk* of the system unit and/or
2. on a *floppy disk* placed in the disk drive.

Saving at least one copy of your file onto a floppy disk in case the file in your PC is deleted is a smart move.

> You must use a floppy disk *formatted* for your computer. If the floppy disk is new, first format it for your PC. If the floppy disk has already been formatted make sure that the format is compatible with your PC. If it is not, re-format it.

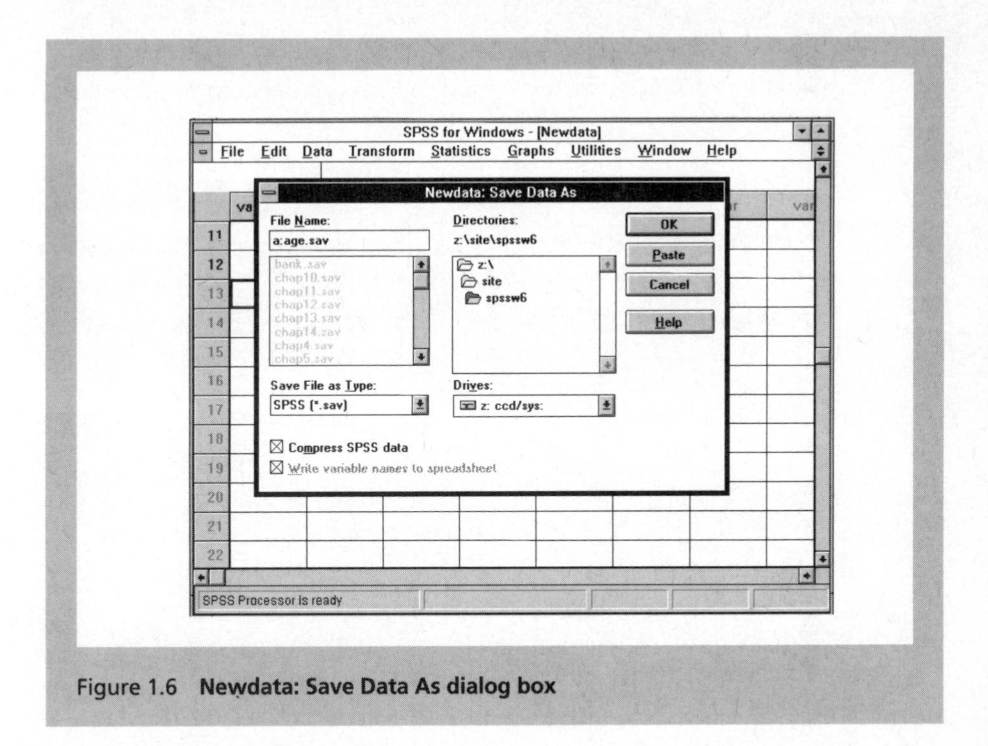

Figure 1.6 **Newdata: Save Data As dialog box**

Data are usually stored in a *system* file. Files are given names which consist of

1. A prefix (or *stem*) of up to eight characters followed by a full stop. The stem name usually refers to the content of the file (such as age in our example).
2. A suffix or *extension* of three characters. The extension name refers to the type of file. The extension name automatically given to system files (i.e. *by default*) is 'sav'.

Consequently we could call our file 'age.sav'. You retrieve the file for later use with its file name.

Perform the following steps to save the data as a system file on a floppy disk in disk drive 'a':

- Move the cursor to the option 'File' on the *menu bar* near the top of the Applications bar and select it to give a *drop-down* menu (Figure 1.5).
- Move the cursor to the 'Save As . . .' option and select it to produce the 'Newdata: Save Data As' dialog box (Figure 1.6).
- Move the cursor to the box under 'File Name:', type the disk drive name followed by a colon 'a:', and the file stem name 'age' so that the file name reads 'a:age.sav'. Typing in this information deletes the asterisk originally present.

- Move the cursor to the button labelled 'OK' and select it. The data has now been saved as a system file named 'age.sav' on the floppy disk in the disk drive called 'a'.

> If you wish to use another drive such as 'c', just type 'c' instead of 'a' so that the complete file name reads 'c:age.sav'.

To retrieve this file from a floppy disk in disk drive 'a', carry out the following steps.

- Move the cursor to the option 'File' on the menu bar near the top of the Applications bar and select it to give a drop-down menu (Figure 1.5).
- Move the cursor to the option 'Open' from the File drop-down menu which produces a second drop-down menu.
- Move the cursor to the option 'Data . . . ' from this second drop-down menu which opens the Open Data File dialog box (Figure 1.7).

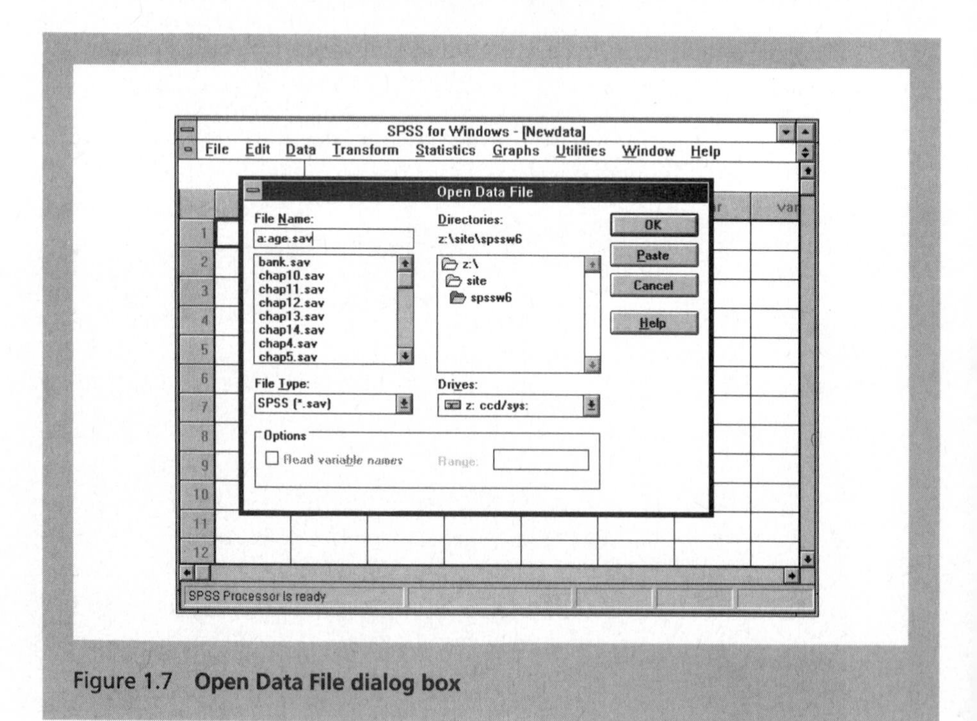

Figure 1.7 **Open Data File dialog box**

- Move the cursor to the box under 'File Name:', type the disk drive name followed by a colon 'a:', and the file stem name 'age' so that the file name reads 'a:age.sav' as shown in Figure 1.7.
- Move the cursor to the button labelled 'OK' and select it. This operation closes the Open Data File dialog box and, after a few moments, opens Newdata with the data in it.

1.8 Clearing Newdata

- If you want to put a new set of data in Newdata (such as the data described in the next chapter), clear the data already entered by moving the cursor on to the File option on the menu bar in the Applications window and select it. This produces a drop-down menu (Figure 1.5).
- Move the cursor onto the New option on this drop-down menu and select it which produces a second drop-down menu.
- Move the cursor onto the Data option on this second drop-down menu which will clear the data in Newdata. You can now type in a new set of data.

1.9 Printing output

- One way of printing the information displayed in the Output window if a printer is appropriately connected to your personal computer is to select the material you want printed.
- Select File which produces a drop-down menu (Figure 1.5).
- Select Print . . . which produces the Print!Output1 dialog box (Figure 1.8).
- Select OK to print out the selected section.

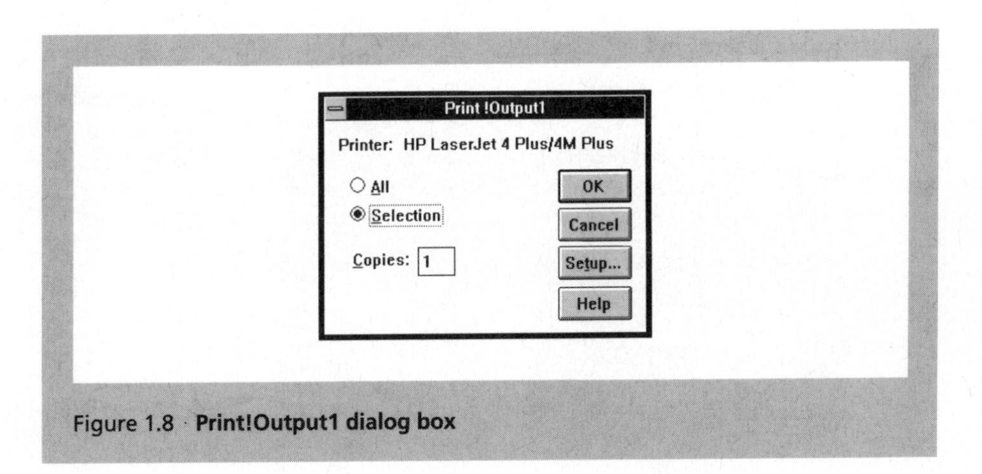

Figure 1.8 · **Print!Output1 dialog box**

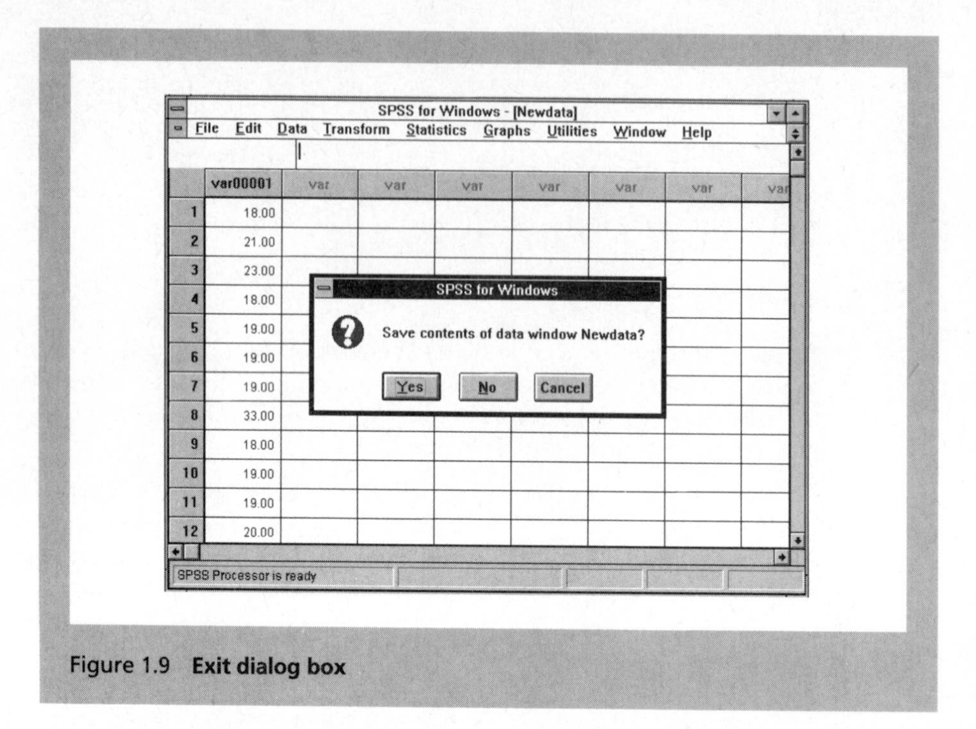

Figure 1.9 **Exit dialog box**

1.10 To leave SPSS

- Move the cursor to the option 'File' on the menu bar in the Applications bar and select it to display the drop-down menu (Figure 1.5).
- Move the cursor to the 'Exit' option (see Figure 1.5) when a small dialog box (Figure 1.9) will be shown.
- As we have already saved the contents of Newdata we select 'No' and we exit from SPSS.

> You cannot break a computer by experimenting with the mouse and keyboard. So novices to computers might care to play for a minute or two to develop their skills. If by any chance things 'lock up' – that is, pressing the keys or mouse has no effect – simply switch off. You will have to start again from scratch, however, if this happens.

1.11 What to do next

Although you may wish to work systematically through this guide, it is feasible to pick out the chapters most relevant to your needs.

Chapter 2

Describing variables

Tables and diagrams

Clear tables and diagrams are a key aspect of statistical analysis and report writing. They are required for virtually every statistical analysis in order to know about the distribution of each of our variables. Computer output usually needs simplifying and clarifying before it is included in reports.

SPSS is generally used to summarise raw data rather than data which have already been summarised such as those shown in Table 2.1 (*ISP* Table 2.1).

In other words, since the data in Table 2.1 are based on 80 people, the data would occupy 80 cells of one column in the Newdata window and each occupation would be coded with a separate number so that Nuns might be coded 1, Nursery Teachers 2 and so on. However, it is possible to carry out certain analyses on summarised data provided that we appropriately weight the categories by the number or frequency of cases in them.

Table 2.1 **Occupational status of participants in the research expressed as frequencies and percentage frequencies**

Occupation	Frequency	Percentage frequency
Nuns	17	21.3
Nursery teachers	3	3.8
Television presenters	23	28.8
Students	20	25.0
Other	17	21.3

2.1 Weighting categories

(Skip this section if you are using raw data.)

Quick summary

Enter code for categories in first column of Newdata

Enter frequency of cases in second column

Data

Weight Cases . . .

Weight cases by

var00002

OK

- We need to give a different numerical code to each of the five categories or groups in Table 2.1. So we will call Nuns 1, Nursery Teachers 2 and so on. Each of the first five cells in the first column of Newdata will contain one of these codes as shown in Figure 2.1.
- The second column will contain the frequency of the five occupational groups.
- Select Data from the menu bar near the top of the screen which produces a drop-down menu (Figure 2.2). We will refer to moving the cursor to a particular

	var00001	var00002	var
1	1.00	17.00	
2	2.00	3.00	
3	3.00	23.00	
4	4.00	20.00	
5	5.00	17.00	
6			

Figure 2.1 **Newdata window containing the code and frequency of the five occupational categories**

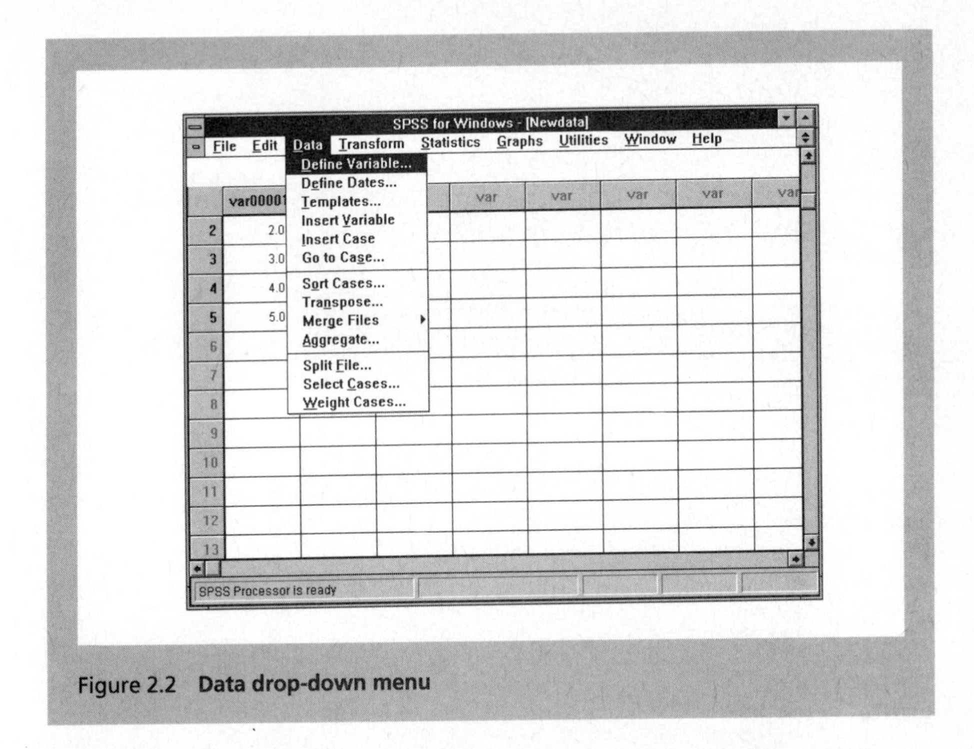

Figure 2.2 Data drop-down menu

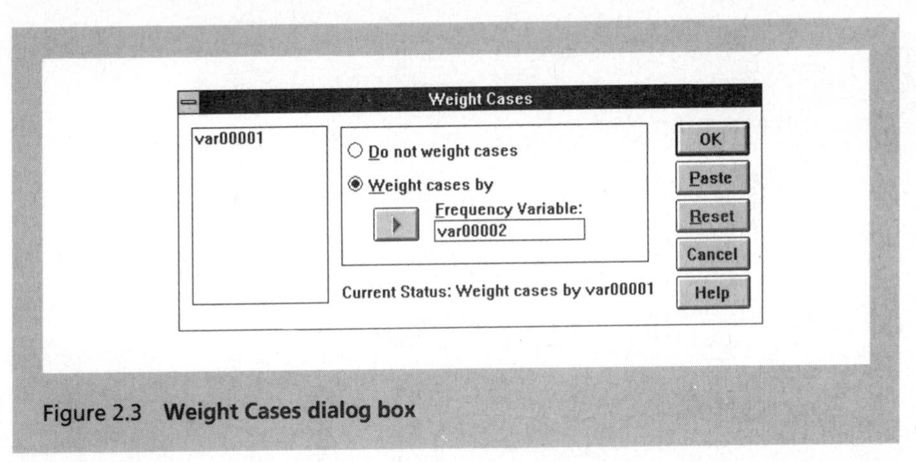

Figure 2.3 Weight Cases dialog box

option and clicking on that option as 'Select', i.e. in this case 'select the Data option' or more simply as 'select Data'.

- Select Weight Cases . . . from this drop-down menu which opens the Weight Cases dialog box (Figure 2.3).
- Select Weight cases by.

- Select 'var00002' and then the ▶ button which puts 'var00002' in the Frequency Variable: text box.
- Select OK which closes the Weight Cases dialog box. The five cells are now weighted by the numbers in the second column.

2.2 Labelling variables and their values

Quick summary

Column to be labelled

Data

Define Variable . . .

Brief name (e.g. Occupat)

Labels . . .

Label (e.g. Occupation)

Value:

Numerical code

Value Label: box

label

Add

Repeat as necessary

Continue

OK

- You may find it helpful to label the variable and each of the different categories of the variable in order to make things clearer. The following procedure does this for the five codes in the first column of Newdata.
- Select the first column (i.e. make sure your cursor or the active cell which is enframed in bold is in the first column).
- Select Data from the menu bar in the Applications window which produces a drop-down menu (Figure 2.2).
- Select Define Variable . . . which opens the Define Variable dialog box (Figure 2.4).

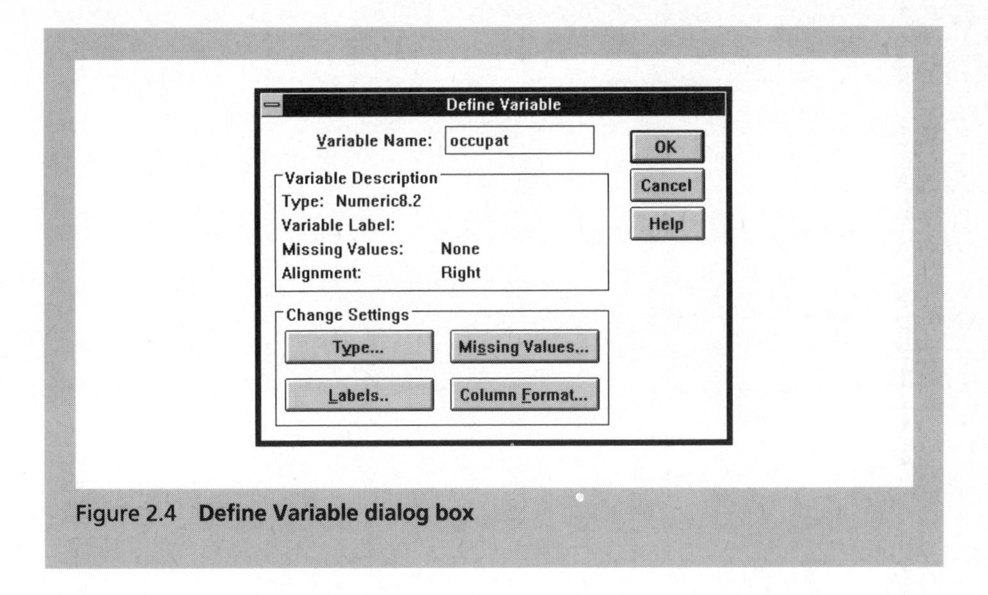

Figure 2.4 **Define Variable dialog box**

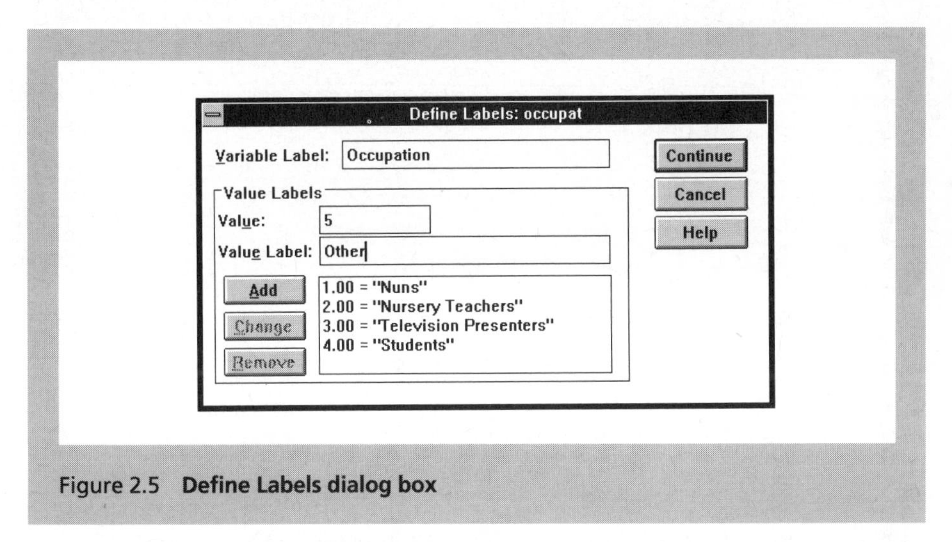

Figure 2.5 **Define Labels dialog box**

- Check that the selected variable is the correct one (i.e. var00001 in our example). Confusing error messages may occur later on if you have not selected the correct variable.
- Variables in Newdata can have a label of up to eight characters so we will call 'var00001' 'occupat' by typing in 'occupat' which will delete the highlighted default name 'var00001' in the Variable Name: box.
- Select Labels . . . which opens the Define Labels dialog box (Figure 2.5).

- We can provide a longer version of the column name which will only be displayed in the output by typing this name in the Variable Label: box. So type in the longer name of Occupation.
- Select Value: in the Value Label section and type 1 in the box beside it.
- Select box beside Value Label: and type Nun in it.
- Select Add which puts 1.00 = "Nuns" in the bottom box.
- Repeat this procedure for the remaining four values and labels as shown in Figure 2.5.
- Select Continue which closes the Define Labels dialog box.
- Select OK which closes the Define Variable dialog box. The first column in Newdata is now labelled 'occupat'.

2.3 Percentage frequencies

Quick summary

Statistics

Summarize

Frequencies . . .

Column variable (e.g. occupat)

OK

- Select Statistics on the menu bar of the Applications window which produces a drop-down menu (Figure 2.6).
- Select Summarize which displays a second drop-down menu which is positioned to the left of the first drop-down menu in Figure 2.6 but which can be positioned to its right depending on the arrangement of the window.
- Select Frequencies . . . which opens the Frequencies dialog box (Figure 2.7).
- Select 'occupat' and the ▶ button which puts 'occupat' in the Variable[s]: text box.
- Select OK which closes the Frequencies dialog box and the Newdata window and which displays the output shown in Table 2.2 in the Output window.

Table 2.2 **Frequency table produced by the Frequency procedure**

OCCUPAT Occupation

Value Label	Value	Frequency	Percent	Valid Percent	Cum Percent
Nuns	1.00	17	21.3	21.3	21.3
Nursery Teachers	2.00	3	3.8	3.8	25.0
Television Presenter	3.00	23	28.8	28.8	53.8
Students	4.00	20	25.0	25.0	78.8
Other	5.00	17	21.3	21.3	100.0
	Total	80	100.0	100.0	

Valid cases 80 Missing cases 0

2.4 Interpreting the output in Table 2.2

- The column name of 'OCCUPAT' is presented together with the longer variable name of 'Occupation'.
- Column 1: The value labels are listed: Nuns, Nursery Teachers, etc.
- Column 2: The numerical values are listed, 1.00, 2.00, etc.
- Column 3: The frequency of each of the categories is presented. Thus there are 17 cases coded as 1.00 which has the value label 'Nuns'.
- Column 4: The percentage of cases in each category for the sample as a whole is displayed. This includes any values that you may have defined as missing values of which there are none in this example. Thus 21.3% of the cases in the sample are coded 1.00 (i.e. Nuns).
- Column 5: The percentage of cases in each category for the sample excluding any missing values is presented. Since there were no values defined as missing in our example, columns 4 and 5 are identical (see Chapter 15).
- Column 6: The cumulative percentage excluding any missing values is found in the final column of numbers. Thus 78.8% of the cases had a value of 4.00 or less.

2.5 Reporting the results in Table 2.2

- Table 2.2 presents most of the main features of these data. It can be used as a model. Extend it if necessary to include more information.
- Be wary of reporting all of the SPSS output since there is a certain amount of 'overkill'. It may be sufficient to report the data shown in Table 2.1.
- Avoid SPSS-specific terms, especially 'valid percent'.

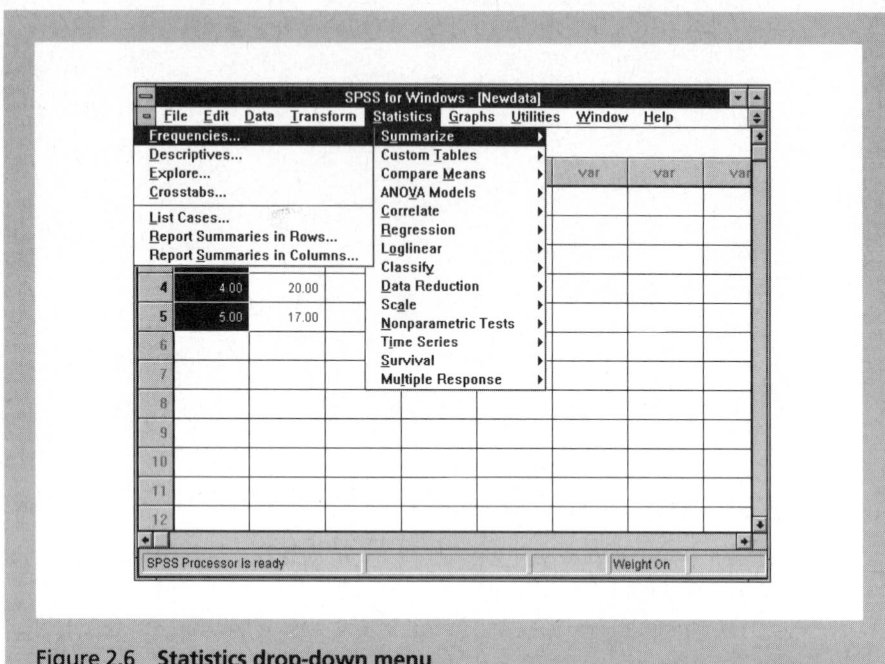

Figure 2.6 **Statistics drop-down menu**

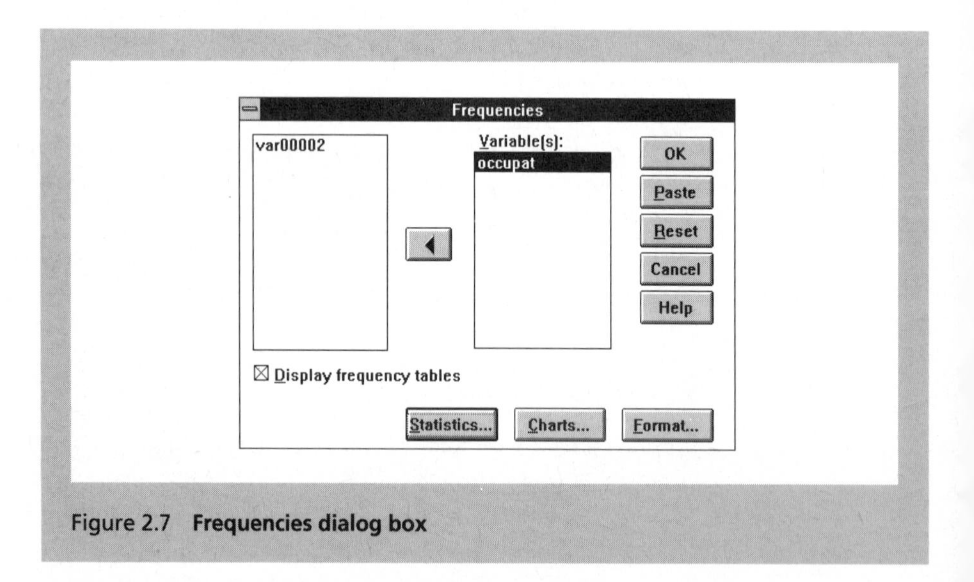

Figure 2.7 **Frequencies dialog box**

2.6 Pie diagram of category data

Quick summary

Graphs

Pie . . .

Define

Column variable (e.g. occupat)

% of cases

OK

- Select Graphs on the menu bar in the Applications window which displays a drop-down menu (Figure 2.8).
- Select Pie . . . which opens the Pie Charts dialog box (Figure 2.9).
- Select Define which opens the Define Pie: Summaries for Groups of Cases dialog box (Figure 2.10).

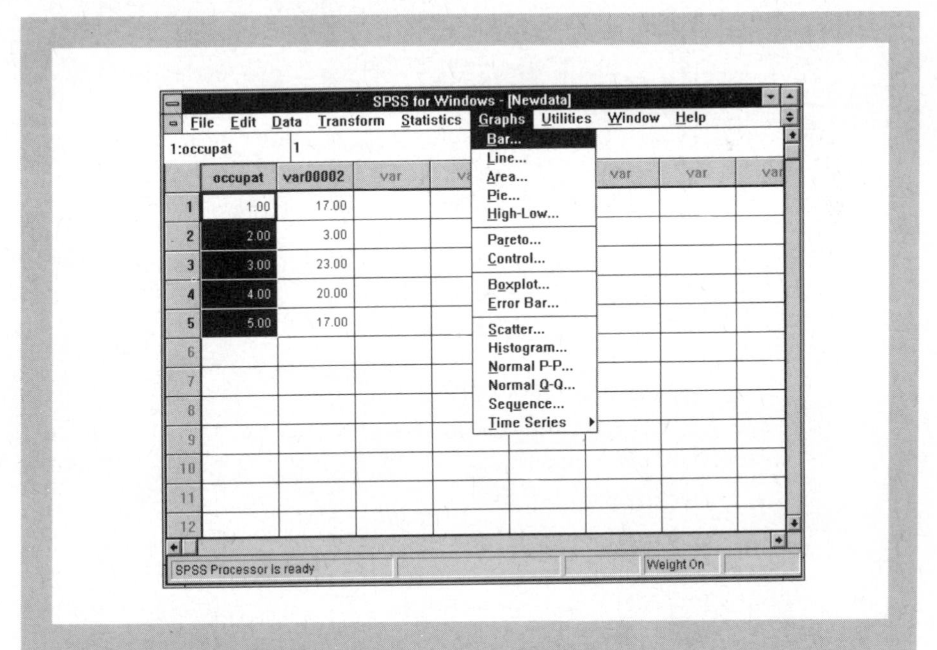

Figure 2.8 **Graphs drop-down menu**

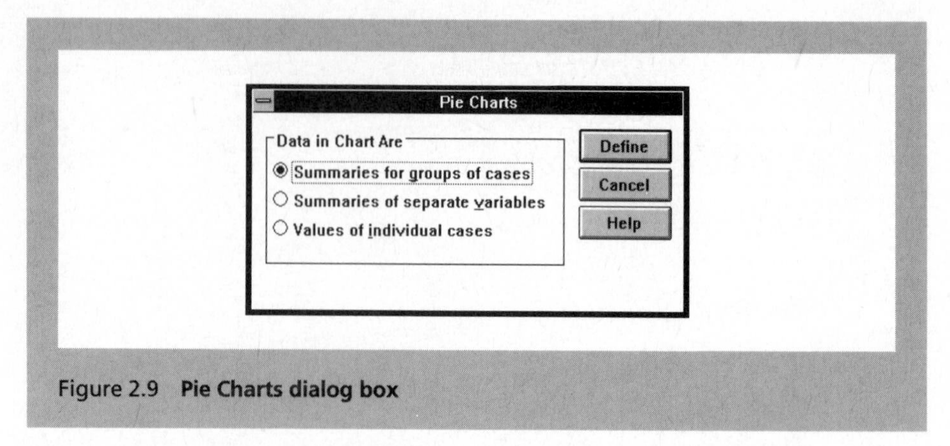

Figure 2.9 **Pie Charts dialog box**

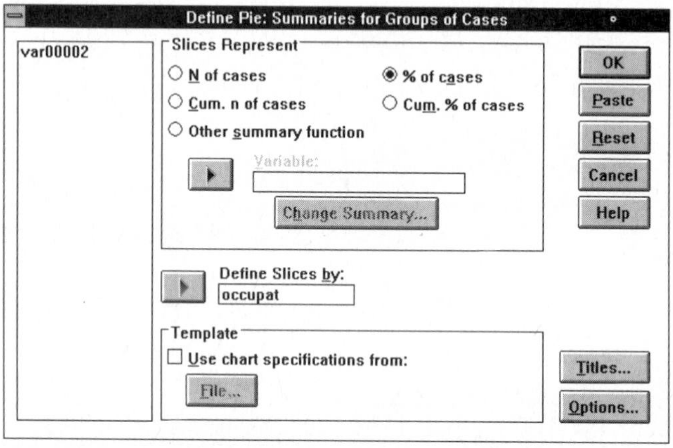

Figure 2.10 **Define Pie: Summaries for Groups of Cases dialog box**

- Select 'occupat' and then the ▶ button which puts 'occupat' in Define Slices by: text box as shown in Figure 2.10.
- Select % of cases.
- Select OK which closes the Define Pie: Summaries for Groups of Cases dialog box and the Newdata window and which displays the pie diagram shown in Figure 2.11 in the Chart Carousel window.
- To return to the Newdata window, select the Window option on the menu bar in the Applications window and select Newdata.

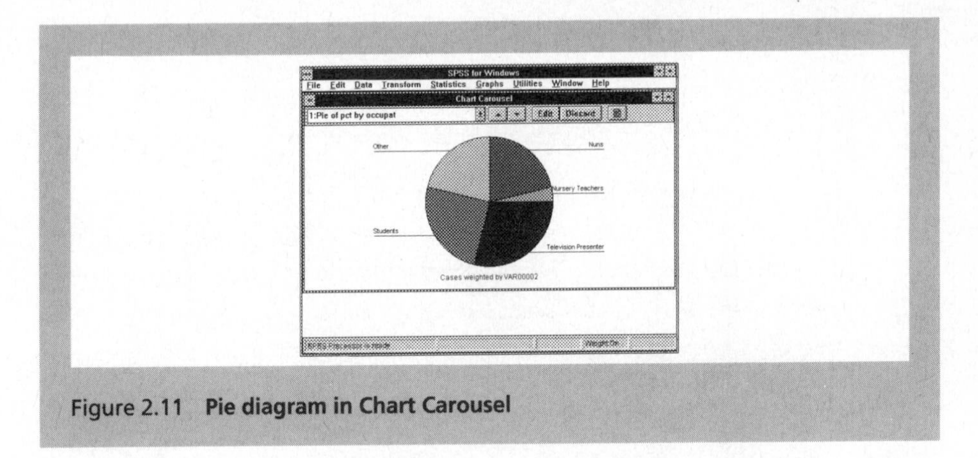

Figure 2.11 **Pie diagram in Chart Carousel**

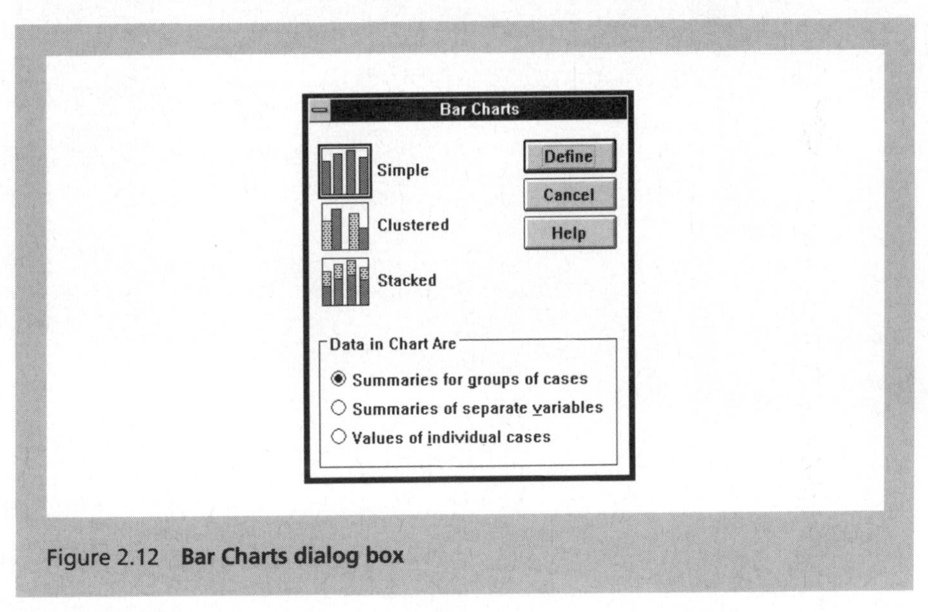

Figure 2.12 **Bar Charts dialog box**

2.7 Bar chart of category data

Quick summary

Graphs

Bar . . .

Define

Column variable (e.g. occupat)

% of cases

OK

- Select Graphs which produces a drop-down menu (Figure 2.8).
- Select Bar . . . which opens the Bar Charts dialog box (Figure 2.12).
- Select Define which produces the Define Simple Bar: Summaries for Groups of Cases dialog box (Figure 2.13).

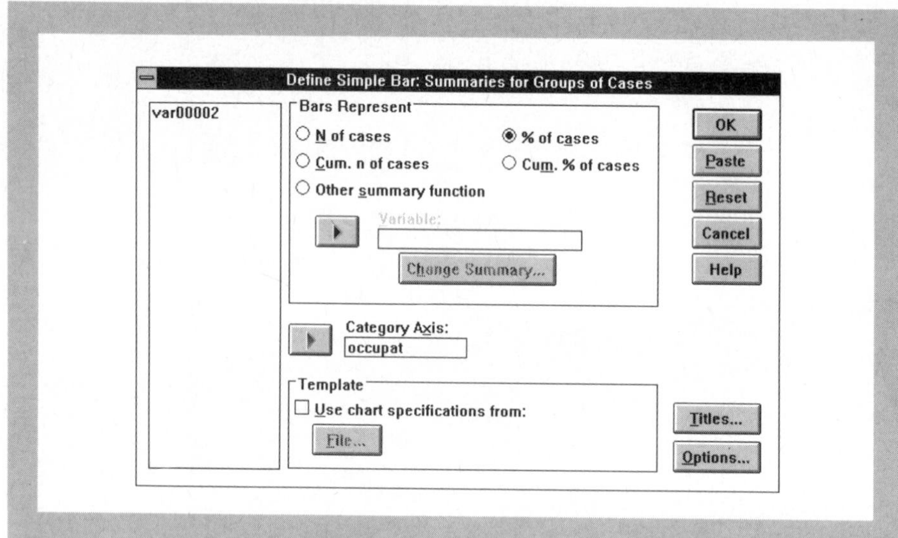

Figure 2.13 **Define Simple Bar: Summaries for Groups of Cases dialog box**

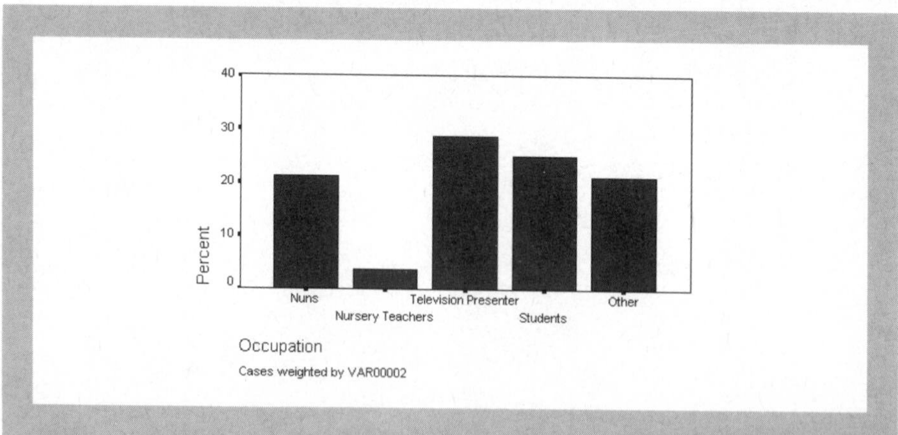

Figure 2.14 **Bar chart of frequency of occupational categories**

- Select 'occupat' and then the ▶ button beside the Category Axis: text box which puts 'occupat' in this box as shown in Figure 2.13.
- Select % of cases.
- Select OK which closes the Define Simple Bar: Summaries for Groups of Cases dialog box and the Newdata window and which displays the bar chart (Figure 2.14) in the Chart Carousel window.

2.8 Histograms

Quick summary

Graphs

Histogram . . .

Column variable (e.g. response)

OK

We will illustrate the drawing up of a histogram with the data in Table 2.3 which shows the distribution of students' attitudes towards statistics. Enter the data in Newdata and weight and label them as described at the beginning of this chapter (we have labelled this variable 'response').

Table 2.3 **Distribution of students' attitudes towards statistics**

Response category	Value	Frequency
Strongly agree	1	17
Agree	2	14
Neither agree nor disagree	3	6
Disagree	4	2
Strongly disagree	5	1

- Select Graphs which produces a drop-down menu (Figure 2.8).
- Select Histogram . . . which opens the Histogram dialog box (Figure 2.15).
- Select 'response' and then the ▶ button which puts 'response' in the Variable: text box as shown in Figure 2.15.
- Select OK which closes the Histogram dialog box and the Newdata window and which displays the histogram presented in Figure 2.16 in the Chart Carousel window.

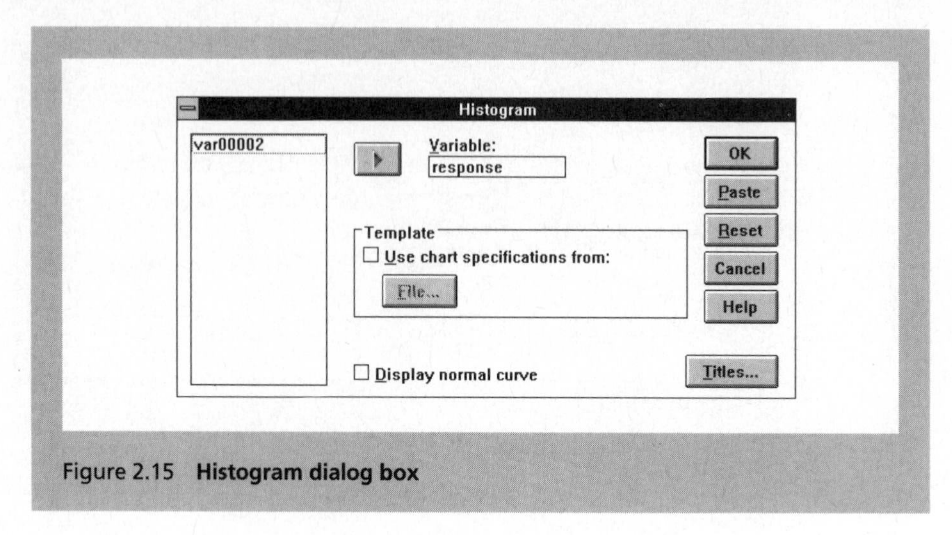

Figure 2.15 **Histogram dialog box**

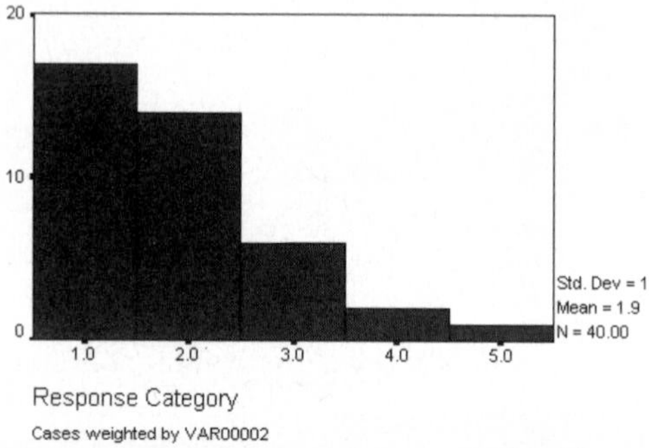

Figure 2.16 **Histogram as displayed in the Chart Carousel window**

Chapter 3

Describing variables numerically

Averages, variation and spread

We will illustrate the computation of the mean, median and mode with the ages of 12 university students shown in Table 3.1 (*ISP* Table 3.7).

Either type the ages in the first column of Newdata or retrieve the file containing them as described in Chapter 1 and repeated below.

Table 3.1 **Twelve numerical scores**

18
21
23
18
19
19
19
33
18
19
19
20

3.1 Retrieving system files

Quick summary

File

Open

Data . . .

File Name:

name of location and file (a:ext.sav)

OK

- Select File from the menu bar in the Applications window which produces a drop-down menu (Figure 1.5).
- Select Open from the File drop-down menu which produces a second drop-down menu.
- Select Data . . . from this second drop-down menu which opens the Open Data File dialog box (Figure 1.7).
- In the File Name: box type in the name of the location of the file (e.g. a), followed by a colon and the file name (age.sav) as shown in Figure 1.7.
- Select OK which closes the Open Data File dialog box and which, after a few moments, opens Newdata with the data in it.

3.2 Descriptive statistics

Quick summary

Statistics

Summarize

Frequencies

Display – OK warning message

'var00001' ▶ *button*

Statistics

Mean, Median and Mode

Continue

OK

- Select Statistics on the menu bar of the Applications window which produces a drop-down menu (Figure 2.6).
- Select Summarize which displays a second drop-down menu (Figure 2.6).
- Select Frequencies . . . which opens the Frequencies dialog box (Figure 2.7).

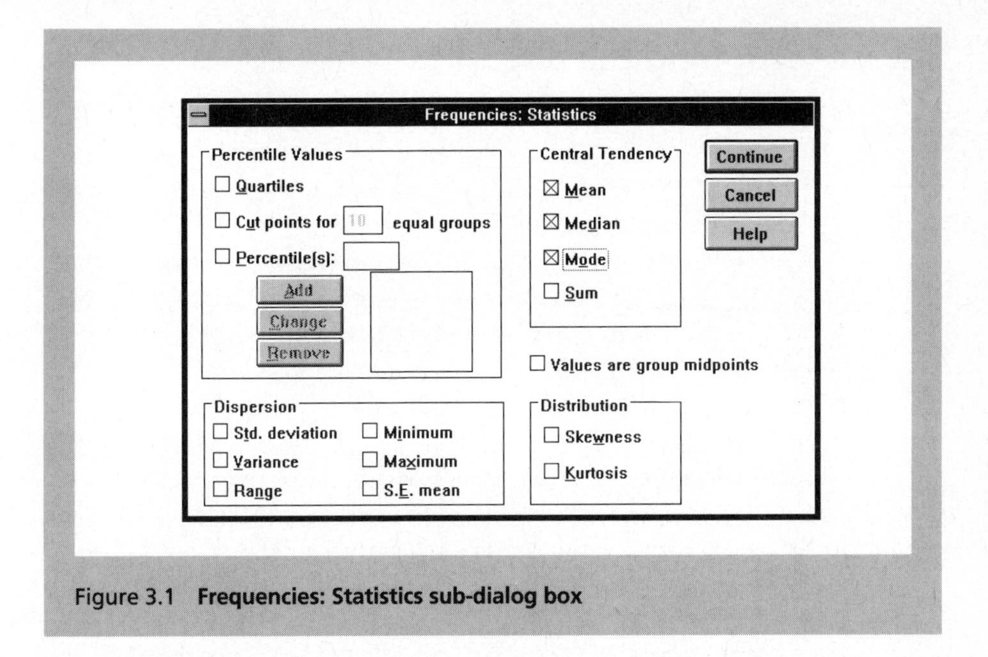

Figure 3.1 **Frequencies: Statistics sub-dialog box**

- Select Display frequency tables which are otherwise automatically produced and which we do not require. There is a query from the computer at this point. Select the OK button.
- Select 'var00001' and the ▶ button which puts 'var00001' in the Variable[s]: text box.
- Select Statistics which opens a second or sub-dialog box called Frequencies: Statistics (Figure 3.1).
- Select Mean, Median and Mode in the box entitled Central Tendency as shown in Figure 3.1. A cross ☒ should appear to indicate the choices you make.
- Select Continue which closes the Frequencies: Statistics sub-dialog box.
- Select OK which closes the Frequencies dialog box and the Newdata window and which displays the output shown in Table 3.2 in the Output window.

Table 3.2 **Mean, median and mode produced by the Frequencies procedure**

```
VAR00001

Mean     20.500        Median  19.000        Mode    19.000

Valid cases  12        Missing cases 0
```

3.3 Interpreting the output in Table 3.2

- The output gives us the following information about VAR00001 which we know to be the ages of a group of university students:
  - (a) The mean age in the group is 20.500 years
  - (b) The median age in the group is 19.000 years
  - (c) The modal age in the group is 19.000 years
- By comparing these three measures of central tendency, it appears that the distribution is not symmetrical. This can be confirmed by looking at Table 3.1.
- The output tells us that there are 12 Valid cases and 0 Missing cases:
  - (a) The Valid cases is the number of ages which have been included in the analysis. In this case it equals the number of scores in Table 1.1.
  - (b) The Missing cases is the number of scores which have been disregarded for the purposes of the analysis. Do not let this worry you, but as long as Missing cases = 0 then there is no problem. (SPSS allows you to identify particular values of a variable as 'missing'. If the computer comes across these for a particular variable, they will be disregarded for analysis purposes. The only circumstance in which this will happen is if you have specifically identified certain values that you wish to be ignored – see Chapter 15.)

3.4 Reporting the output in Table 3.2

- Two decimal places are more than enough for most psychological data. Most measurement in psychology is approximate and the use of several decimal places tends to imply an unwarranted degree of precision.
- For the median and mode, it is probably less confusing if you do not report values as 19.00 but as 19. However, if the decimal places are anything other than .00 then this should be reported, since it indicates that the median or mode is estimated and does not correspond to any actual scores: see Table 3.3.

Table 3.3 **Mean, median and mode of age**

	Ages of students (N = 12)
Mean	20.50
Median	19
Mode	19

3.5 Other features

You will see from Figure 3.1 that in the Window there are many additional statistical values which may be calculated. You should have little difficulty in obtaining these by adapting the steps already described:

- Percentiles
- Quartiles
- Mean
- Median
- Mode
- Sum
- Skewness
- Kurtosis
- Standard deviation (estimate)
- Variance (estimate)
- Range
- Minimum (score)
- Maximum (score)
- Standard error (S.E. mean)

Chapter 4

Shapes of distributions of scores

It is important to know about the shape of the distribution of scores on a variable. Ideally for parametric statistics, a distribution should be symmetrical and normal (bell-shaped). Histograms, for example, will reveal marked discrepancies from this ideal.

We will compute a frequency table and histogram of the extraversion scores of 50 airline pilots shown in Table 4.1 (*ISP* Table 4.1).

Table 4.1 **Extraversion scores of 50 airline pilots**

3	5	5	4	4	5	5	3	5	2
1	2	5	3	2	1	2	3	3	3
4	2	5	5	4	2	4	5	1	5
5	3	3	4	1	4	2	5	1	2
3	2	5	4	2	1	2	3	4	1

4.1 Frequency tables

Quick summary

Enter the data (Chapter 1) or retrieve file

Statistics

Summarize

Frequencies

Select 'var00001' ▶ *button*

OK

- As described in Chapter 1 enter the extraversion scores in the first column of the Newdata window.
- Save the data, as described in Chapter 1, in a system file called 'ext.sav'.
- Select Statistics which produces a drop-down menu (Figure 2.6).
- Select Summarize which displays a second drop-down menu (Figure 2.6).
- Select Frequencies . . . which opens the Frequencies dialog box (Figure 2.7).
- Select 'var00001' (if you have only one variable this will already be highlighted for you) and then the ▶ button which puts 'var00001' in the Variable[s] text: box.
- Select OK which closes the Frequencies dialog box and the Newdata window and which displays the output shown in Table 4.2 in the Output window.

Table 4.2 **Frequency table of extraversion scores of 50 airline pilots as produced by Frequencies**

VAR00001

Value Label	Value	Frequency	Percent	Valid Percent	Cum Percent
	1.00	7	14.0	14.0	14.0
	2.00	11	22.0	22.0	36.0
	3.00	10	20.0	20.0	56.0
	4.00	9	18.0	18.0	74.0
	5.00	13	26.0	26.0	100.0
	Total	50	100.0	100.0	

Valid cases 50 Missing cases 0

4.2 Interpreting the output in Table 4.2

- Column 1: The five categories of extraversion scores are listed in the left-most column of numbers. They are 1.00, 2.00, 3.00, 4.00 and 5.00.
- Column 2: The frequency of each of these categories is to be found. Thus there are 10 cases with an extraversion value of 3.00.
- Column 3: The percentage of cases in each category for the sample as a whole. This includes any values that you may have defined as missing values. Thus 20% of the cases in the sample are in the 3.00 category.
- Column 4: The percentage of cases in each category for the sample excluding any missing values. Since there were no values defined as missing in our example, columns 3 and 4 are identical.

- Column 5: The cumulative percentage excluding any missing values. Thus 74% of the cases had a value of 4.00 or less.
- There is nothing entered under Value Label since we have not defined any value labels.

4.3 Reporting the output in Table 4.2

Table 4.3 shows one style of reporting the output. Notice that we omitted some of the confusion of detail in Table 4.2. Statistical tables and diagrams need to clarify the results.

Table 4.3 **One style of reporting output in Table 4.2**

Extraversion score	Frequency	Percentage frequency	Cumulative percentage frequency
1	7	14.0	14.0
2	11	22.0	36.0
3	10	20.0	56.0
4	9	18.0	74.0
5	13	26.0	100.0

4.4 Histograms

Quick summary

Enter the data (Chapter 1) or retrieve file

Graphs

Histogram . . .

Select 'var00001' ▶ button

OK

- Select Graphs which produces a drop-down menu (Figure 2.8).
- Select Histogram . . . which displays the Histogram dialog box (Figure 2.15).
- Select 'var00001' (which will already be highlighted for you if you only have one variable) and then the ▶ button which puts 'var00001' in the Variable text box.

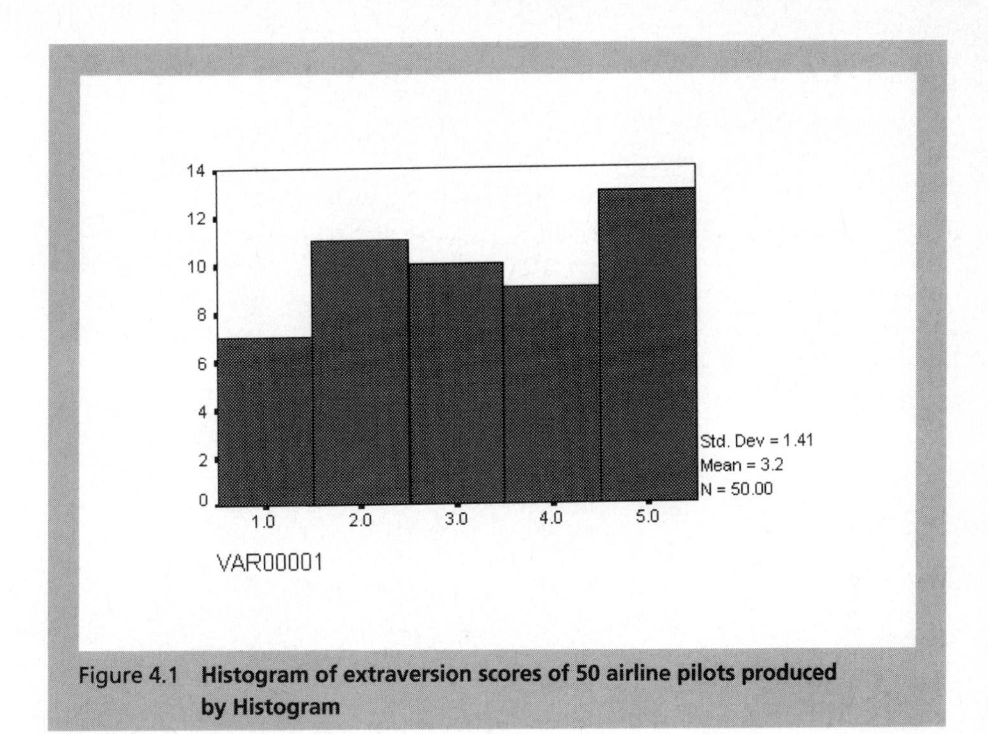

Figure 4.1 **Histogram of extraversion scores of 50 airline pilots produced by Histogram**

- Select OK which closes the Histogram dialog box and the Newdata window and which displays the output shown in Figure 4.1 in the Chart Carousel window.

4.5 Interpreting the output in Figure 4.1

- The horizontal axis of this histogram gives the values 1.0, 2.0, 3.0, 4.0 and 5.0. These are the five different values of extraversion scores in Table 4.1.
- The vertical axis gives frequencies of individuals or cases from 0 to 14 in steps of 2.
- The heights of the bars tell you how frequent each extraversion score is in the data.
- Std. Dev = 1.41 is the value of the standard deviation. This is discussed again in Chapter 5 of this book.
- The average of the 50 extraversion scores is given by Mean = 3.2.
- N = 50.00 means that the number of scores = 50.

4.6 Reporting the output in Figure 4.1

- Although Figure 4.1 may be easily interpreted by you, it is not suitable for inclusion unmodified in your reports.
- The vertical axis needs to be labelled 'f' or 'frequencies'.
- The horizontal axis needs to be labelled 'extraversion scores'.

Chapter 5

Standard deviation

The standard unit of measurement in statistics

Standard deviation is an index of how much scores deviate (differ) 'on average' from the average of the set of scores of which they are members. It can also be used in order to turn scores on very different variables into 'standard scores' which are easily compared.

We shall illustrate the computation of the standard deviation and z-scores with the nine age scores ($N = 9$) shown in Table 5.1 (based on *ISP* Table 5.1).

Table 5.1 **Data for the calculation of standard deviation**

Age
20
25
19
35
19
17
15
30
27

5.1 Standard deviation

Quick summary

Statistics

Summarize

Descriptives

Select variables ▶

Options

Std. deviation

Continue

OK

- Select Statistics which produces a drop-down menu (Figure 2.6).
- Select Summarize which displays a second drop-down menu (also Figure 2.6).

either

- Select Frequencies . . . and follow the procedure described in Chapter 3, selecting Std. deviation in the Frequencies: Statistics sub-dialog box (Figure 3.1)

or

- Select Descriptives . . . which opens the Descriptives dialog box (Figure 5.1).
- Select 'var00001' and then the ▶ button which puts 'var00001' in the Variable[s]: text box.
- Select Options . . . which opens the Descriptives: Options sub-dialog box (Figure 5.2).

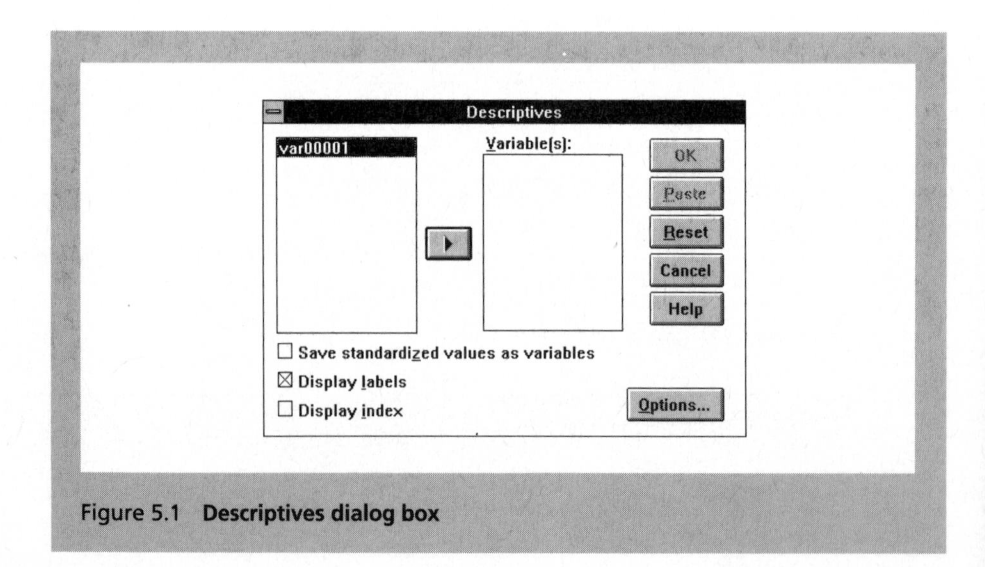

Figure 5.1 Descriptives dialog box

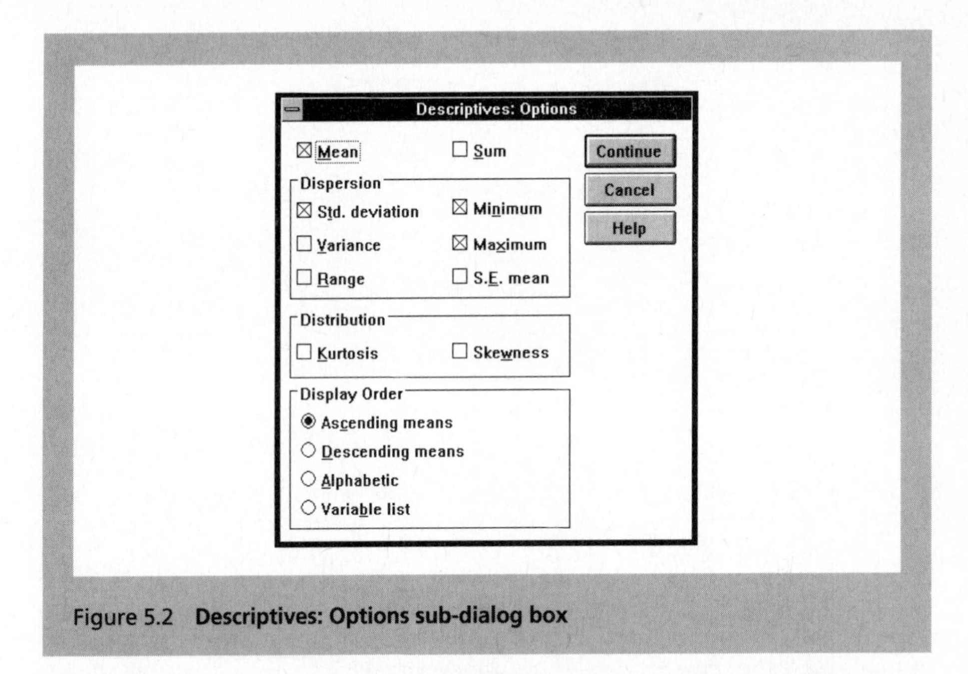

Figure 5.2 **Descriptives: Options sub-dialog box**

- Select Std. deviation and de-select Mean, Minimum and Maximum by clicking off these options (if you only want standard deviations). Normally you will find the mean and minimum plus maximum useful.
- Select Continue which closes the Descriptives: Define sub-dialog box.
- Select OK which closes the Descriptives dialog box and Newdata window and which displays the output shown in Table 5.2 in the Output window.

Table 5.2 **Standard deviation of nine ages as produced by Descriptives**

Number of valid observations (listwise) = 9.00

Variable	Std Dev	Valid N
VAR00001	6.65	9

5.2 *Z*-scores

Quick summary

Statistics

Summarize

Descriptives

Select 'Save standardized values as variables'

Select variables ▶

Options

Std. deviation

Continue

OK

This involves just one extra step:

- To obtain *z*-scores we simply select 'Save standardized values as variables' in the Descriptives dialog box (Figure 5.1) which will put the *z*-scores in the next column in Newdata entitled 'zvar0000' as shown in Figure 5.3.

In other words, for each of your scores the *z*-score has been calculated and placed in a separate column in your data window. This new 'variable' will be lost unless you save your data before ending the session.

5.3 Interpreting the output in Table 5.2

- The standard deviation of the nine ages (VAR00001) is 6.65. This is the 'average' amount by which each score deviates from the mean score in the set.

1:zvar0000 -.450987616801697

	var00001	zvar0000	var
1	20.00	-.45099	
2	25.00	.30066	
3	19.00	-.60132	
4	35.00	1.80395	
5	19.00	-.60132	
6	17.00	-.90198	
7	15.00	-1.20263	
8	30.00	1.05230	
9	27.00	.60132	

Figure 5.3 ***z*-Scores produced by Descriptives and displayed in Newdata**

Standard deviation is quite a complex concept, so refer to *ISP* Chapter 5 if you have any difficulties.

- The standard deviation is based on nine valid cases. In other words, the computer has not found any values of age which you have previously defined as 'missing' or to be ignored (Chapter 15).

5.4 Reporting the output in Table 5.2

- The standard deviation of just one variable can easily be mentioned in the text of your report: 'It was found that the standard deviation of age was 6.65 years (N = 9).'
- However, it is more likely that you would wish to record the standard deviation alongside other statistics such as the mean and range, as illustrated in Table 5.3. You probably wish to include these statistics for other numerical score variables on which you have data.

Table 5.3 **Sample size, mean, range and standard deviations of age, IQ and verbal fluency**

	N	Mean	Range	Standard deviation
Age	9	23.00	20.00	6.65
IQ	9	122.17	17.42	14.38
Verbal fluency	9	18.23	4.91	2.36

5.5 Other features

Descriptives contains a number of statistical calculations which are easily selected:

- Mean
- Sum
- Standard deviation (estimate)
- Range
- Minimum (score)
- Maximum (score)
- Standard error (S.E. mean)
- Kurtosis
- Skewness

Chapter 6

Relationships between two or more variables

Diagrams and tables

Much research is about the relationship between two or more variables. This chapter explains how to draw tables and diagrams which illustrate the relationships between variables.

We will illustrate the drawing-up of a crosstabulation table and compound bar chart with the data shown in Table 6.1 (*ISP* Table 6.4). This shows the number of men and women who have or have not been previously hospitalised.

Table 6.1 **Crosstabulation table of sex against hospitalisation**

	Male	**Female**
Previously hospitalised	$f = 20$	$f = 25$
Not previously hospitalised	$f = 30$	$f = 14$

6.1 Weighting data

Quick summary

Enter code for rows in first column

Enter code for columns in second column

Enter frequencies in third column

Data

Weight Cases . . .

Weight cases by

var00003

OK

- We need to identify each of the four cells in Table 6.1. The rows of the table represent whether or not participants have been hospitalised while the columns represent the sex of the participants. We will then weight each of the four cells of the table by the number of cases in them.
- The first column of Figure 6.1 contains the code for whether participants have been previously hospitalised (1) or not (2).
- The second column contains the code for males (1) and females (2).
- The third column has the frequency of people in each of these four cells.
- Select Data from the menu bar in the Applications window which produces a drop-down menu (Figure 2.2).
- Select Weight Cases . . . from this drop-down menu which opens the Weight Cases dialog box (Figure 2.3).
- Select Weight cases by.
- Select 'var00003' and then the ▶ button which puts 'var00003' in the Frequency Variable: text box.
- Select OK which closes the Weight Cases dialog box. The four cells are now weighted by the numbers in the third column.

	var00001	var00002	var00003	va
1	1.00	1.00	20.00	
2	1.00	2.00	25.00	
3	2.00	1.00	30.00	
4	2.00	2.00	14.00	
5				

Figure 6.1 Weighted data in Newdata

6.2 Labelling variables and their values

Quick summary

Column to be labelled

Data

Define Variable . . .

Brief name (e.g. hospital)

Labels . . .

Label (Hospitalisation)

Value:

Numerical code

Value Label: box

label

Add

Repeat as necessary

Continue

OK

- Select the first column.
- Select Data from the menu bar in the Applications window which produces a drop-down menu (Figure 2.2).
- Select Define Variable . . . which opens the Define Variable dialog box (Figure 2.4).
- Type 'Hospital' in the Variable Name: box.
- Select Labels . . . which opens the Define Labels dialog box (Figure 2.5).
- Type Hospitalisation in the Variable Label: box.
- Select Value: in the Value Label section and type 1 in the box beside it.
- Select box beside Value Label: and in it type Hospitalised.
- Select Add which puts 1.00 = 'Hospitalised' in the bottom box.
- Select Value: in the Value Label section and type 2 in the box beside it.
- Select box beside Value Label: and in it type Not hospitalised.

- Select Add which puts 2.00 = 'Not hospitalised' in the bottom box.
- Select Continue which closes the Define Labels dialog box.
- Select OK which closes the Define Variable dialog box. The first column in Newdata is now labelled 'hospital'.
- Select the second column.
- Select Data from the menu bar in the Applications window which produces the drop-down menu shown in Figure 2.2.
- Select Define Variable . . . which opens the Define Variable dialog box (Figure 2.4).
- Type 'Gender' in the Variable Name: box.
- Select Labels . . . which opens the Define Labels dialog box (Figure 2.5).
- Type Gender in the Variable Label: box.
- Select Value: in the Value Label section and type 1 in the box beside it.
- Select box beside Value Label: and in it type Male.
- Select Add which puts 1.00 = 'Male' in the bottom box.
- Select Value: in the Value Label section and type 2 in the box beside it.
- Select box beside Value Label: and in it type Female.
- Select Add which puts 2.00 = 'Female' in the bottom box.
- Select Continue which closes the Define Labels dialog box.
- Select OK which closes the Define Variable dialog box. The first column in Newdata is now labelled 'gender'.

6.3 Crosstabulation with frequencies

Quick summary

Statistics

Custom Tables

Basic Tables . . .

Row variable (e.g. hospital)

▶

Column variable (e.g. gender)

OK

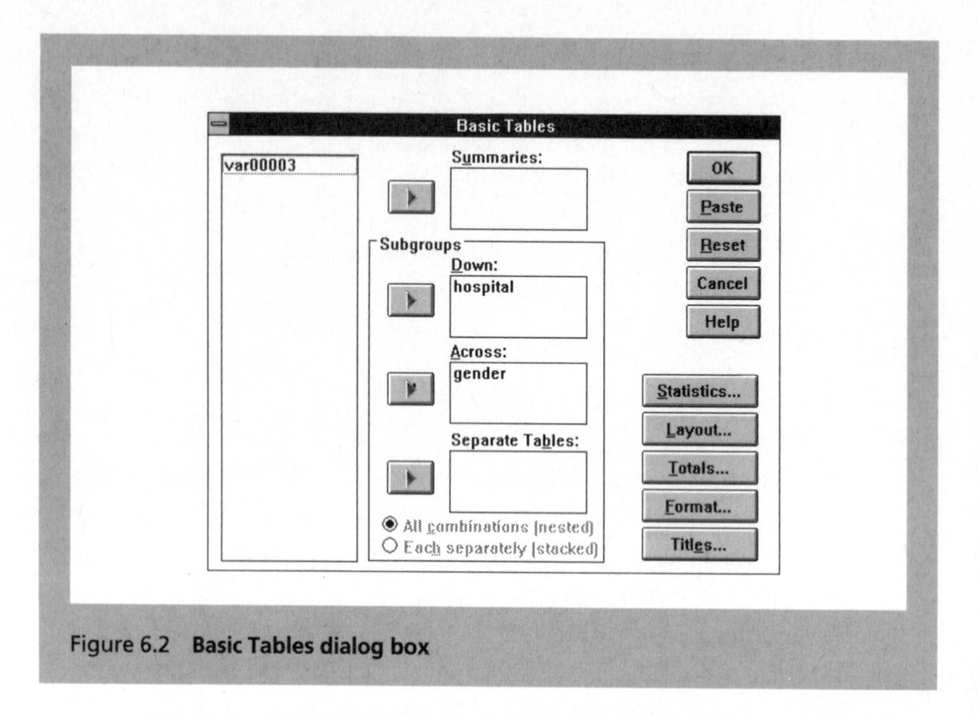

Figure 6.2 **Basic Tables dialog box**

The following three steps in **bold type** should be followed to get to the Basic Tables dialog box for all of the analyses in this section.

- **Select Statistics on the menu bar of the Applications window which produces a drop-down menu (Figure 2.6).**
- **Select Custom Tables on this drop-down menu which opens a second drop-down menu.**
- **Select Basic Tables . . . on this second drop-down menu which opens the Basic Tables dialog box (Figure 6.2: note that this dialog box contains entries for the next two steps).**
- If you want to put the hospitalisation variable in the rows of the table, select 'hospital' and then the ▶ button beside the Down: box which puts 'hospital' in this box.
- If you want to put the gender variable in the columns of the table, select 'gender' and then the ▶ button beside the Across: box which puts 'gender' in this box.
- Select OK which closes the Basic Tables dialog box and the Newdata window and which presents the output shown in Table 6.2 in the Output window.
- To display the frequencies in each cell as a percentage of the total, **follow the three steps in bold at the start of this section**, then select Statistics . . . in the Basic Tables dialog box which opens the Basic Tables: Statistics dialog box (Figure 6.3).

Table 6.2 **Crosstabulation table of gender against hospitalisation produced by Basic Tables**

	Gender	
	Male	Female
Hospitalisation		
Hospitalised	20	25
Not hospitalised	30	14

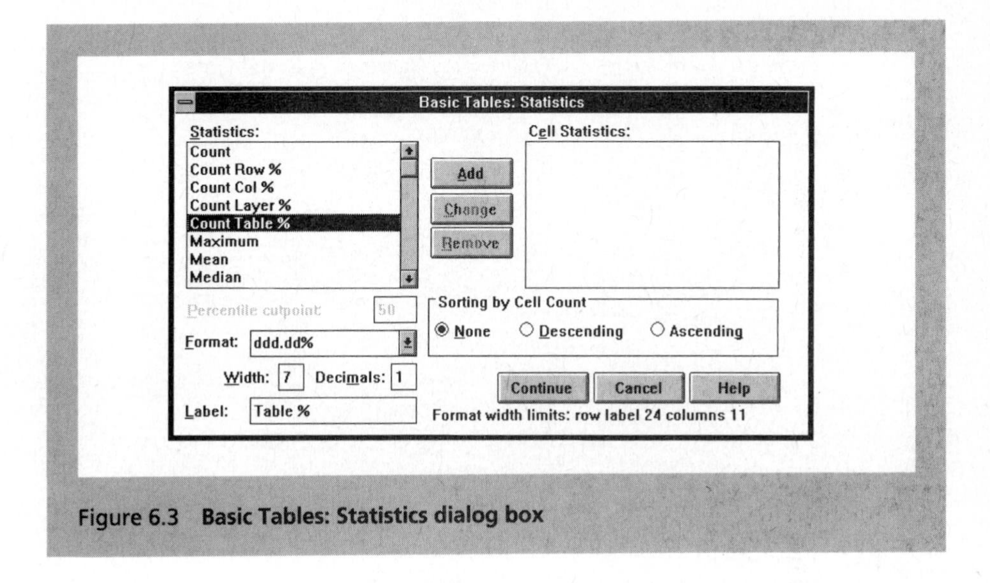

Figure 6.3 **Basic Tables: Statistics dialog box**

- Select Count Table % and then Add which puts this term in the Cell Statistics: box.
- Select Continue which closes the Basic Tables: Statistics dialog box.
- Select OK which closes the Basic Tables dialog box and the Newdata window and which presents the output shown in Table 6.3 in the Output window.
- To display the frequencies in each cell as a percentage of the column total, **follow the three steps in bold at the start of this section**, then select Statistics . . . in the Basic Tables dialog box which opens the Basic Tables: Statistics dialog box (Figure 6.3).
- Select Count Col % and then Add which puts this term in the Cell Statistics: box. (If Count Table % is still in the Cell Statistics: box from the previous operation, then select it and then Add which will put it back in the Statistics: box. If you do not do this you will obtain both the table and the column percentages.)

Table 6.3 **Crosstabulation table with frequencies as a percentage of the total**

	Gender	
	Male	Female
	Table %	Table %
Hospitalisation		
Hospitalised	22.5%	28.1%
Not hospitalised	33.7%	15.7%

Table 6.4 **Crosstabulation table with frequencies as percentage of the column total**

	Gender	
	Male	Female
	Col %	Col %
Hospitalisation		
Hospitalised	40.0%	64.1%
Not hospitalised	60.0%	35.9%

- Select Continue which closes the Basic Tables: Statistics dialog box.
- Select OK which closes the Basic Tables dialog box and the Newdata window and which presents the output shown in Table 6.4 in the Output window.

6.4 Compound (stacked) percentage bar chart

Quick summary

Graphs

Bar . . .

Stacked

Define

Axis variable (e.g. hospital)

Stacked variable (e.g. gender)

OK

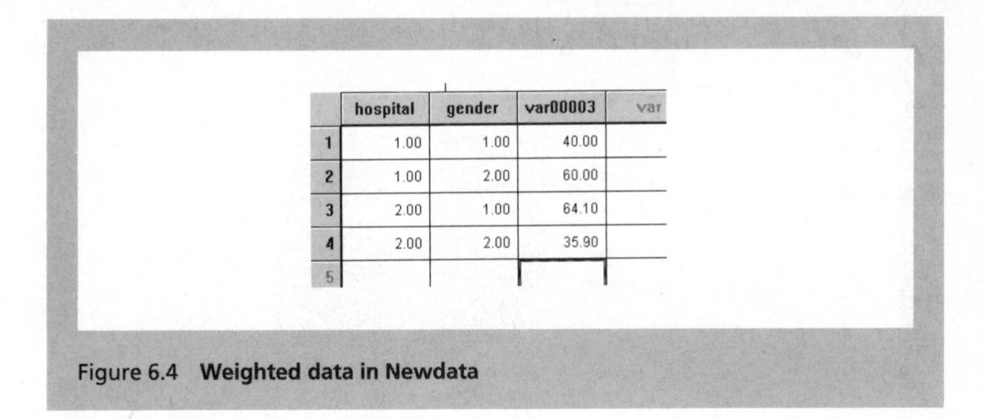

	hospital	gender	var00003	var
1	1.00	1.00	40.00	
2	1.00	2.00	60.00	
3	2.00	1.00	64.10	
4	2.00	2.00	35.90	
5				

Figure 6.4 **Weighted data in Newdata**

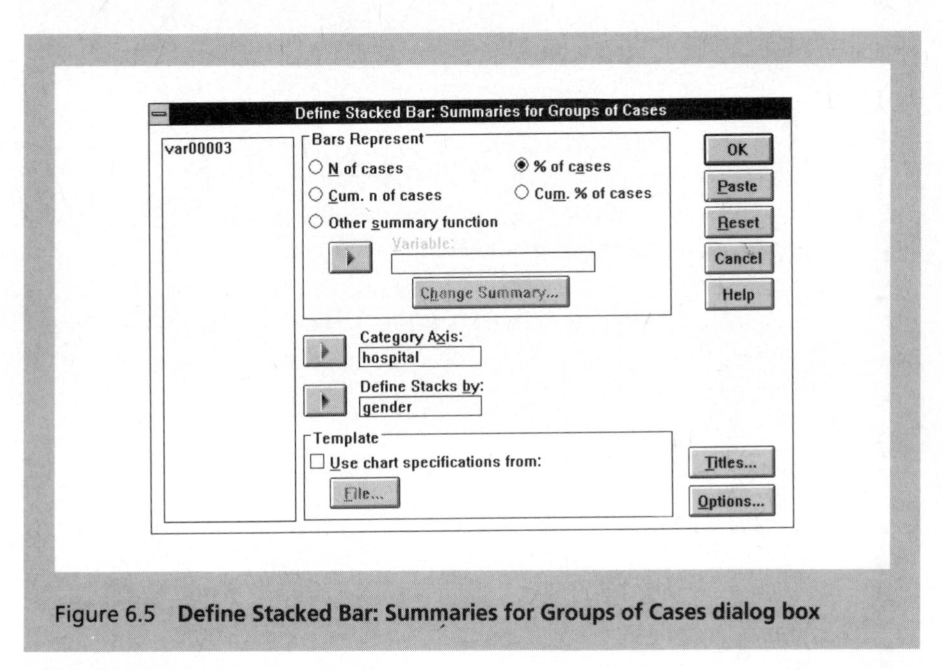

Figure 6.5 **Define Stacked Bar: Summaries for Groups of Cases dialog box**

- SPSS does not directly produce a compound (stacked) percentage bar chart in which the bars represent 100%. To obtain this kind of chart indirectly you need to enter the percentage figures for the two bars as shown in Figure 6.4 and weight them.
- Select Graphs which produces a drop-down menu (Figure 2.8).
- Select Bar . . . which opens the Bar Charts dialog box (Figure 2.12).
- Select Stacked.
- Select Define which produces the Define Stacked Bar: Summaries for Groups of Cases dialog box (Figure 6.5).

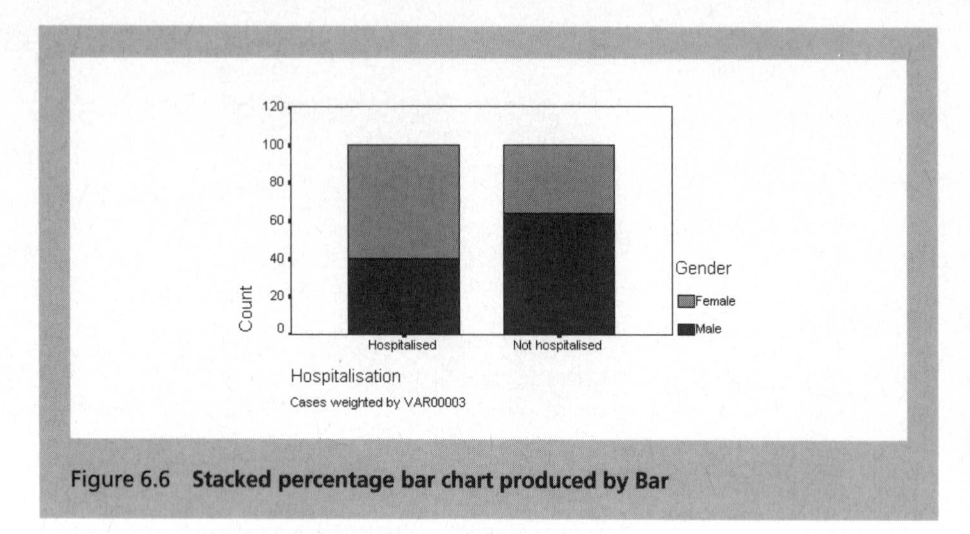

Figure 6.6 **Stacked percentage bar chart produced by Bar**

- Select 'hospital' and then the ▶ button beside the Category Axis: text box which puts 'hospital' in this box as shown in Figure 6.5.
- Select 'gender' and then the ▶ button beside the Define Stacks by: box which puts 'gender' in this box as shown in Figure 6.5.
- Select OK which closes the Define Stacked Bar: Summaries for Groups of Cases dialog box and the Newdata window and which displays the stacked bar chart shown in Figure 6.6 in the Chart Carousel window. Note that the label Count on the vertical axis refers to percent.

6.5 Compound histogram (clustered bar chart)

Quick summary

Graphs

Bar . . .

Clustered

Define

Axis variable (e.g. hospital)

Clustered variable (e.g. gender)

% of cases

OK

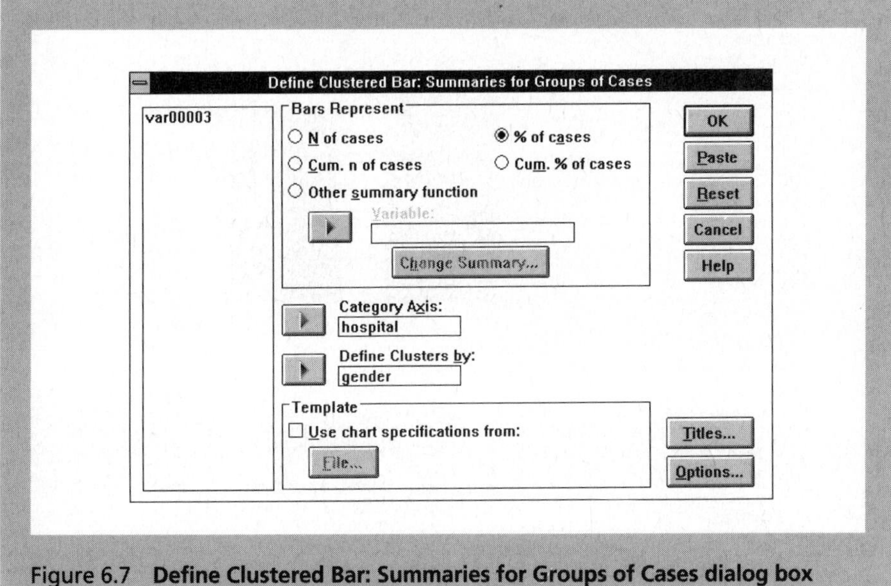

Figure 6.7 **Define Clustered Bar: Summaries for Groups of Cases dialog box**

- To display this stacked bar chart as a compound histogram or clustered bar chart, select Graphs and then Bar . . . , select Clustered in the Bar Charts dialog box (Figure 2.12) which opens the Define Clustered Bar: Summaries for Groups of Cases dialog box (Figure 6.7).
- Select 'hospital' and then the ▶ button beside the Category Axis: text box which puts 'hospital' in this box (Figure 6.7).
- Select 'gender' and then the ▶ button beside the Define Clusters by: box which puts 'gender' in this box (Figure 6.7).
- Select % of cases.
- Select OK which closes the Define Clustered Bar: Summaries for Groups of Cases dialog box and the Newdata window and which displays the clustered bar chart shown in Figure 6.8 in the Chart Carousel window.

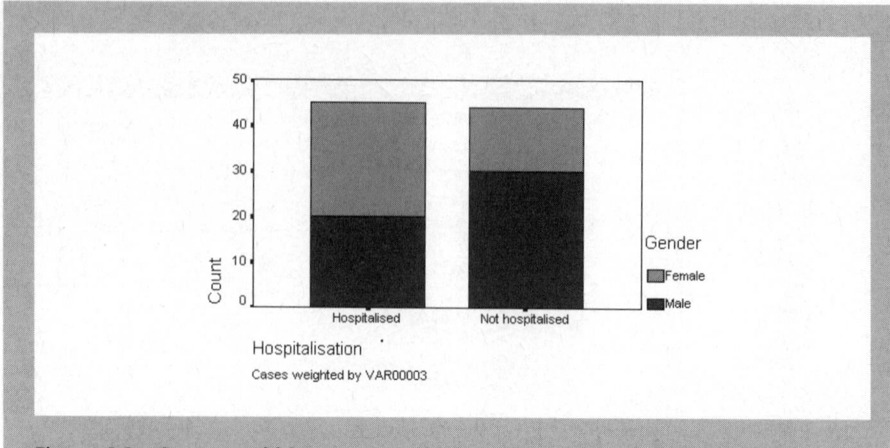

Figure 6.8 **Compound histogram or clustered bar chart produced by Bar**

Chapter 7

Correlation coefficients

Pearson's correlation and Spearman's Rho

The correlation coefficient is a numerical index which describes how closely related two variables are and whether it is a positive relationship (both variables increase numerically) or a negative relationship (scores increase on one variable as they decline on the other).

We will illustrate the computation of Pearson's correlation, a scatter diagram and Spearman's Rho for the data in Table 7.1 (*ISP* Table 7.1) which gives scores for the musical ability and mathematical ability of 10 children. We begin by typing in the music scores in the first column of Newdata and the mathematics scores in the second column (Figure 7.1). We then proceed to analyse the relationship between these two sets of scores.

Table 7.1 **Scores on musical ability and mathematical ability for 10 children**

Music score	Mathematics score
2	8
6	3
4	9
5	7
7	2
7	3
2	9
3	8
5	6
4	7

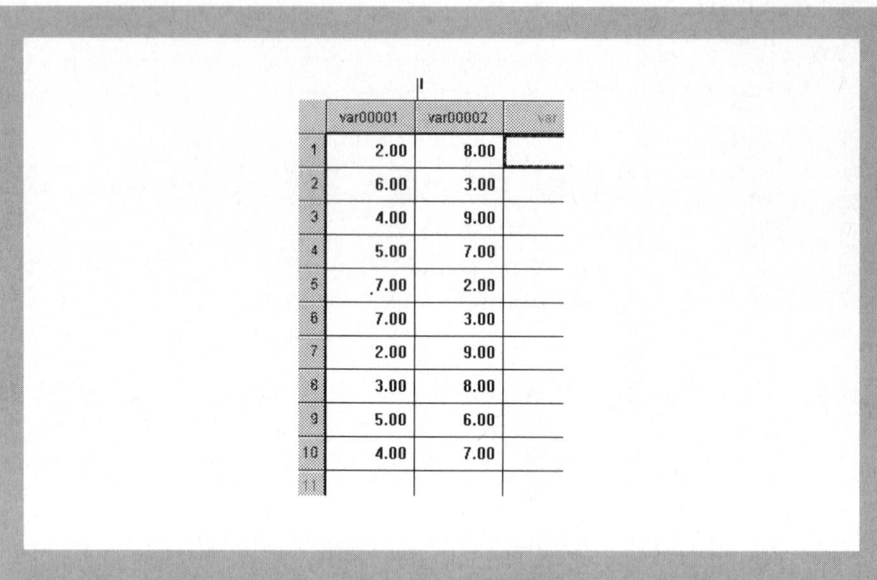

	var00001	var00002
1	2.00	8.00
2	6.00	3.00
3	4.00	9.00
4	5.00	7.00
5	7.00	2.00
6	7.00	3.00
7	2.00	9.00
8	3.00	8.00
9	5.00	6.00
10	4.00	7.00

Figure 7.1 **Music and mathematics scores in Newdata**

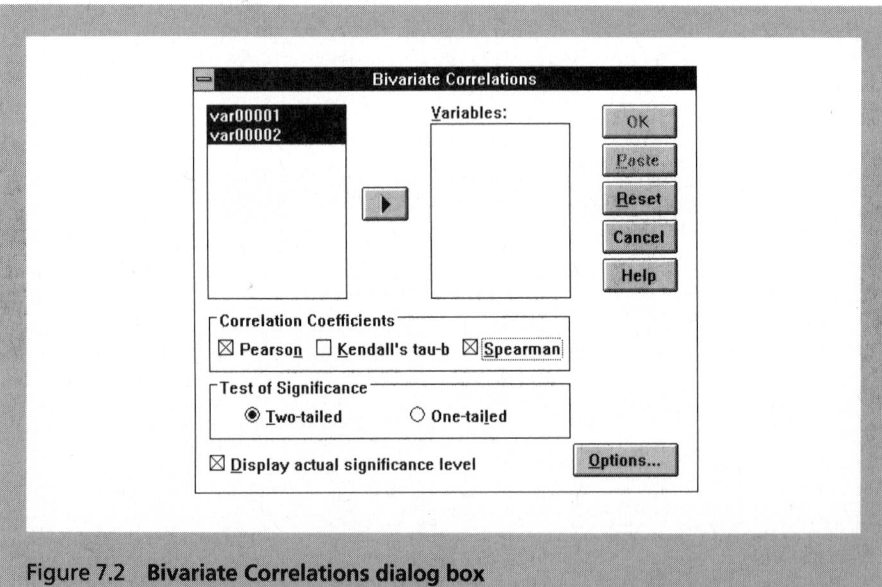

Figure 7.2 **Bivariate Correlations dialog box**

7.1 Pearson's correlation

Quick summary

Statistics

Correlate

Bivariate

Select variables ▶

OK

- Select Statistics from the menu bar in the Applications window which produces a drop-down menu (Figure 2.6).
- Select Correlate from the drop-down menu which reveals a smaller drop-down menu.
- Select Bivariate from this drop-down menu which opens the Bivariate Correlations dialog box (Figure 7.2).
- Select 'var00001' and 'var00002' and then the ▶ button which puts 'var00001' and 'var00002' in the Variables: text box. You can either select the two variables by two separate operations or drag the highlight down over the second variable using the mouse. Click-on to the highlight and move down to cover the second variable before releasing the mouse button.

Table 7.2 **Pearson's correlation produced by Correlate**

```
-- Correlation Coefficients --

                VAR00001       VAR00002

VAR00001         1.0000         -.8998
                (   10)        (   10)
                P= .           P= .000

VAR00002         -.8998         1.0000
                (   10)        (   10)
                P= .000        P= .

(Coefficient / (Cases) / 2-tailed Significance)

'' . '' is printed if a coefficient cannot be computed
```

- The Pearson option has already been pre-selected (i.e. it is a default option), so if only Pearson's correlation is required select OK which closes the Bivariate Correlations dialog box and the Newdata window and produces the output shown in Table 7.2 (p.55) in the Output window.

7.2 Interpreting the output in Table 7.2

- The variables on which the correlation was carried out are given both in the columns and in the rows. We have just two variables so a 2 × 2 correlation matrix is generated.
- The print-out gives indications of how to read the entries in the table `(Coefficient / (Cases) / 2-tailed Significance)`
- The correlation of VAR00001 with VAR00002 is –0.8998.
- Ten pairs of scores were used to obtain the correlation coefficient `(   10)`.
- The exact significance level is given to three decimal places `(P= .000)`. This is a very significant level.
- Correlations are displayed in a matrix. The diagonal of this matrix (from top left to bottom right) consists of the variable correlated with itself which obviously gives a perfect correlation of 1.0000. No significance level is given for this value as it never varies `(P= .  )`.
- The values of the correlations are symmetrical around the diagonal from top right to bottom left in the matrix.

7.3 Reporting the output in Table 7.2

- The correlation between musical ability and mathematical ability is –0.8998. It is usual to round correlations to two decimal places, which would make it –0.90. This is more than precise enough for most psychological measurements.
- The exact significance level to three decimal places is 0.000. This means that the significance level is less than 0.001. We would suggest that you do not use a string of zeros as these confuse people. Always change the final zero to a 1. This means that the significance level can be reported as being $p<0.001$.
- It is customary to present the degrees of freedom rather than the number of cases when presenting correlations. The degrees of freedom are the number of cases minus 2 which makes them 8 for this correlation. There is nothing wrong with reporting the number of cases instead.
- In a report, we would write 'There is a significant negative relationship between musical ability and mathematical ability (r = –0.90, df = 8, $p<0.001$). Children with more musical ability have lower mathematical ability.' Significance of the correlation coefficient is discussed in more detail in the textbook (*ISP*, Chapter 10).

7.4 Spearman's Rho

Quick summary

Statistics

Correlate

Bivariate

Select Spearman and de-select Pearson

Select variables ▶

OK

To correlate the scores in ranks rather than as raw scores, one simply makes a different choice in the Bivariate Correlations box:

- To produce Spearman's correlation by itself, select Spearman in the Bivariate Correlations dialog box shown in Figure 7.1 and de-select Pearson.
- Select OK which closes the Bivariate Correlations dialog box and the New-data window and which produces the output shown in Table 7.3 in the Output window.

Table 7.3 **Spearman's correlation produced by Correlate**

```
---SPEARMAN  CORRELATION  COEFFICIENTS---

VAR00002        -.8944
               N(  10)
               Sig .000

               VAR00001

(Coefficient / (Cases) / 2-tailed Significance)
'' . '' is printed if a coefficient cannot be computed
```

7.5 Interpreting the output in Table 7.3

- The Spearman's instruction does not print out a matrix for just two variables.
- Spearman's correlation between the ranks for musical ability and mathematical ability is –0.8944.

- The number of cases on which that correlation was based is 10 and is given by `N(  10)` in the table.
- The exact significance level is given to three decimal places as `Sig .000` but this is best handled by changing the final 0 to 1.
- The degrees of freedom are the number of cases minus 2 which makes them 8.

7.6 Reporting the output in Table 7.3

- The correlation reported to two decimal places is –0.89.
- The probability of achieving this correlation by chance is less than 0.001 (i.e. $p<0.001$).
- We would report this in the following way: 'There is a statistically significant negative correlation between musical ability and mathematical ability (rho = –0.89, df = 8, $p<0.001$). Those with the highest musical ability tend to be those with the lowest mathematical ability and vice versa.'

7.7 Scatter diagram

Quick summary

Graphs

Scatter

Define

Put one variable into the X-axis ▶ and another into the Y-axis ▶

Select OK

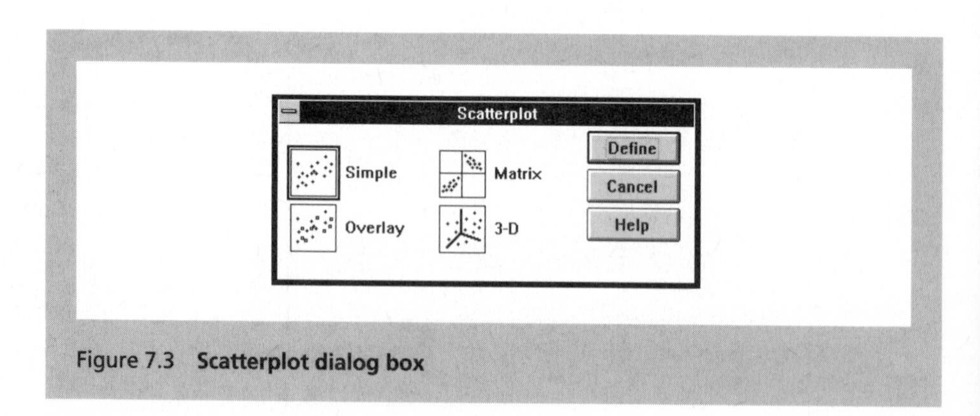

Figure 7.3 **Scatterplot dialog box**

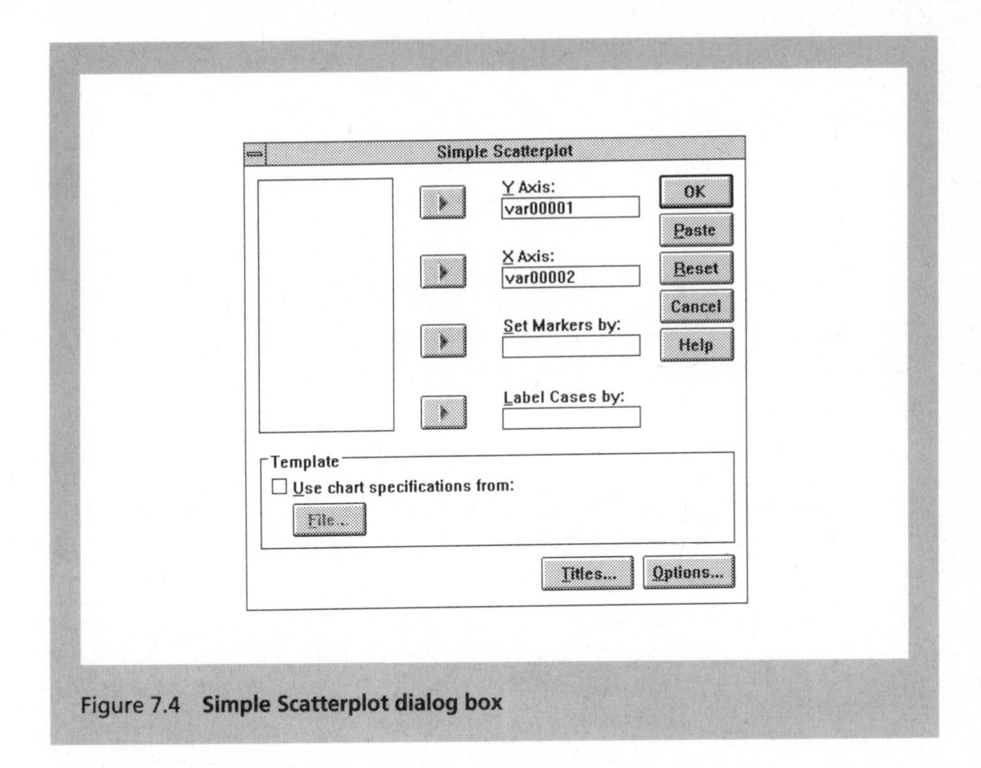

Figure 7.4 **Simple Scatterplot dialog box**

- Select the Graphs option on the menu bar in the Applications window which displays a drop-down menu (Figure 2.8).
- Select Scatter . . . which opens the Scatterplot dialog box (Figure 7.3).
- Since the Simple option is the default and so has already been pre-selected as indicated by the bold square frame surrounding it, select Define which opens the Simple Scatterplot dialog box (Figure 7.4).
- With a correlation it does not really matter which variable represents the horizontal or *X*-axis (the abscissa) and which variable the vertical or *Y*-axis (the ordinate). We will specify the *Y*-axis as being musical ability ('var00001') and the *X*-axis as being mathematical ability ('var00002'), so select 'var00001' and the ▶ button beside the Y Axis: box which puts 'var00001' into this box as shown in Figure 7.4.
- Select 'var00002' and the ▶ button next to the X Axis: box which puts 'var00002' into this box as shown in Figure 7.4.
- Select OK which closes the Simple Scatterplot dialog box and the Newdata window and which displays the scatter diagram shown in Figure 7.5 in the Chart Carousel window.

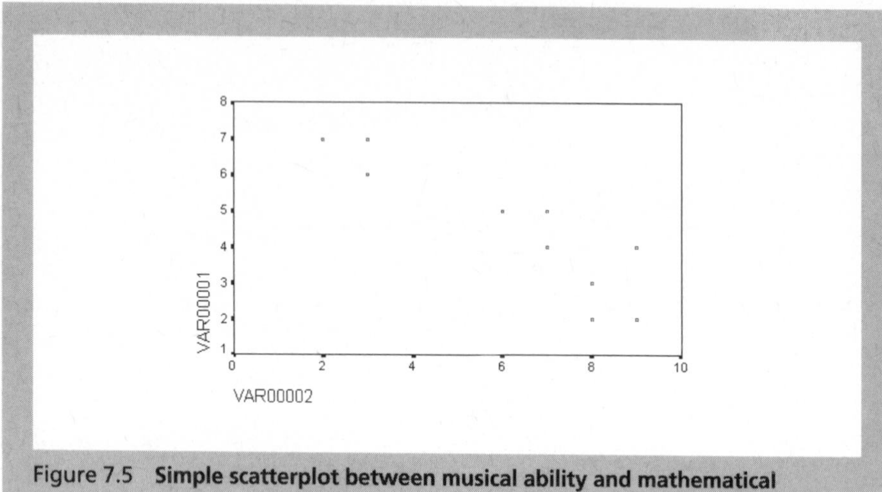

Figure 7.5 **Simple scatterplot between musical ability and mathematical ability produced by Scatter**

7.8 Interpreting Figure 7.5

- The scatter of points is relatively narrow, indicating that the correlation is high.
- The slope of the scatter lies in a relatively straight line, indicating that it is a linear rather than a curvilinear relationship.
- If the relationship is curvilinear, then Pearson's or Spearman's correlation coefficients may be misleading.
- This line moves from the upper left to the lower right which signifies a negative correlation.

7.9 Reporting Figure 7.5

- You should never report a correlation coefficient without examining the scattergram for problems such as curved relationships or outliers (*ISP* Chapter 7).
- In a student project it should always be possible to include graphs of this sort. Unfortunately, journal articles and books tend to be restricted in the numbers they include because of economies of space and cost.
- We would write of Figure 7.5: 'A scattergram of the relationship between mathematical ability and musical ability was examined. There was no evidence of a curvilinear relationship or the undue influence of outliers.'

The Bivariate Correlations procedure allows you also to obtain Kendall's tau and one-tailed tests of significance. There are a variety of scatterplots available. These can be edited in many ways in the Chart Carousel window by selecting Edit. Explore these.

Chapter 8

Regression

Prediction with precision

Regression allows you to predict scores on a variable if you know the relationship between two variables. It can be used on much the same data as the correlation coefficient but is less commonly used because of problems of comparability between values of different sets of variables.

We will illustrate the computation of simple regression and a regression plot with the data in Table 8.1 (*ISP* Table 7.1) which gives scores for the musical ability and mathematical ability of 10 children. These data are identical to those used in the previous chapter on correlation. In this way, you may find it easier to appreciate the differences between regression and correlation.

The music scores ('var00001'), which are in the first column of Newdata, are the criterion or the dependent variable while the mathematics scores ('var00002'), which are in the second column, are the predictor or independent variable. With

Table 8.1 **Scores on musical ability and mathematical ability for 10 children**

Music score	Mathematics score
2	8
6	3
4	9
5	7
7	2
7	3
2	9
3	8
5	6
4	7

regression, it is essential to make the criterion or dependent variable the vertical axis (*Y*-axis) of a scatterplot, and the predictor or independent variable the horizontal axis (*X*-axis).

8.1 Simple regression equations

Quick summary

Statistics

Regression

Linear

Select ▶ variable for Y-axis; select ▶ variable for X-axis

OK

- Select Statistics from the menu bar in the Applications window which produces a drop-down menu (Figure 2.6).
- Select Regression from the drop-down menu which reveals a smaller drop-down menu.
- Select Linear . . . from this drop-down menu which opens the Linear Regression dialog box (Figure 8.1: note this dialog box contains entries for the next two steps).

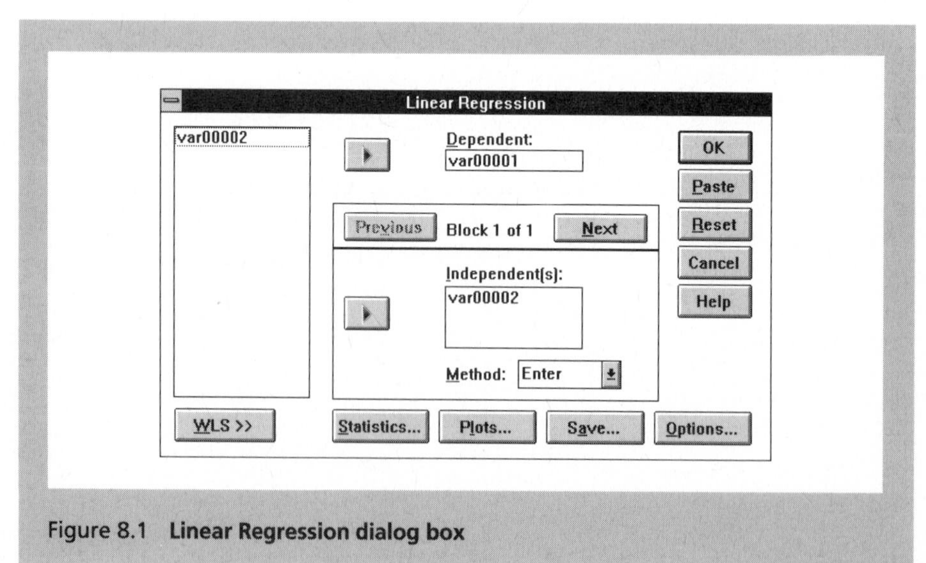

Figure 8.1 **Linear Regression dialog box**

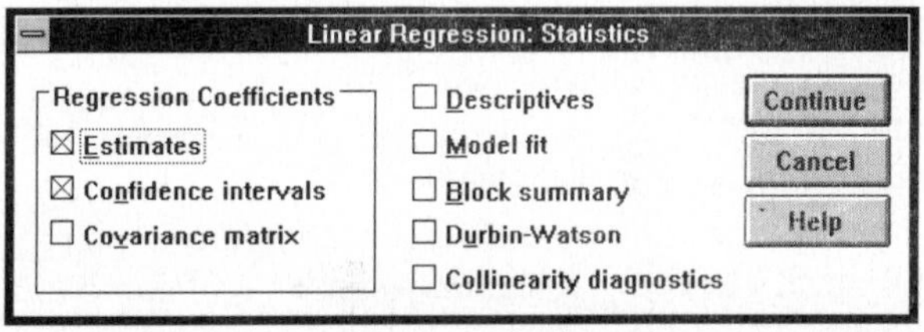

Linear Regression: Statistics

Regression Coefficients
☒ Estimates
☒ Confidence intervals
☐ Covariance matrix

☐ Descriptives
☐ Model fit
☐ Block summary
☐ Durbin-Watson
☐ Collinearity diagnostics

Continue
Cancel
Help

Figure 8.2 **Linear Regression: Statistics dialog box**

- Select 'var00001' and then the ▶ button beside the Dependent: box which puts 'var00001' in this box.
- Select 'var00002' and then the ▶ button beside the Independent[s]: box which puts 'var00002' in this box.
- Select Statistics . . . which opens the Linear Regression: Statistics dialog box (Figure 8.2).
- Select Confidence intervals and de-select Model fit.
- Select Continue which closes the Linear Regression: Statistics dialog box.
- Select OK which closes the Linear Regression dialog box and the Newdata window and which produces the output shown in Table 8.2 in the Output window.

Table 8.2 **Linear regression output**

```
          * * * * M U L T I P L E   R E G R E S S I O N * * * *
Listwise Deletion of Missing Data
Equation Number 1      Dependent Variable..      VAR00001
Block Number 1. Method: Enter      VAR00002
Variable(s) Entered on Step Number
1..      VAR00002
   ----------- Variables in the Equation -----------
Variable             B       SE B  95% Confdnce Intrvl B      Beta
VAR00002      -.633117   .108566  -.883471    -.382763  -.899755
(Constant)    8.425325   .725040  6.753381    10.097268
----------- in -----------
Variable             T     Sig T
VAR00002        -5.832     .0004
(Constant)      11.620     .0000
End Block Number     1     All requested variables entered.
```

8.2 Interpreting the output in Table 8.2

In simple regression involving two variables, it is conventional to report the regression equation as a slope (*a*) and an intercept (*b*) as explained in *ISP* (Chapter 8). SPSS does not quite follow this terminology but all of the relevant information is in Table 8.2. Unfortunately, at this stage the SPSS output is far more complex and detailed than the statistical sophistication of most students. The key basic elements of the output are highlighted in **bold type** in the following list:

- **B is the slope. The slope of the regression line is called the unstandardised regression coefficient in SPSS. The unstandardised regression coefficient between 'var00001' and 'var00002' is displayed under B in the Variables in the Equation section and is –0.633117 which rounded to two decimal places is –0.63.**
- The 95% confidence interval for this coefficient ranges from –0.88 (–0.883471) to –0.38 (–0.382763). Since the regression is based on a sample and not the population, there is always a risk that the population regression coefficient is not the same as that in the population. The 95% confidence interval gives the range of regression slopes within which you can be 95% sure the population slope will lie.
- **The intercept (*a*) is referred to as the constant in SPSS. The intercept is presented as the (Constant) and is 8.425325 which rounded to two decimal places is 8.43. It is the point at which the regression line cuts the vertical (*Y*) axis.**
- The 95% confidence interval for the intercept is `6.753381` to `10.097268`. This means that, based on your sample, the intercept of the population is 95% likely to lie in the range of 6.75 to 10.10.
- The column headed Beta gives a value of `–.899755`. This is actually the Pearson correlation between the two variables. In other words, if you turn your scores into standard scores (*z*-scores) the slope of the regression and the correlation coefficient are the same thing.

8.3 Regression scatterplot

Quick summary

Graphs

Scatter

Define

Select ▶ *variable for* Y-*axis; select* ▶ *variable for* X-*axis*

OK

Chart

Edit (on Chart Carousel window)

Options

Total

OK

It is generally advisable to inspect a scattergram of your two variables when doing regression. This involves the steps involved in plotting a scattergram as described in Chapter 7.

- Select the Graphs option on the menu bar in the Applications window which displays a drop-down menu (Figure 2.8).
- Select Scatter . . . which opens the Scatterplot dialog box (Figure 7.2).
- Since the Simple option is the default and so has already been pre-selected as indicated by the bold square frame surrounding it, select Define which opens the Simple Scatterplot dialog box (Figure 7.3).

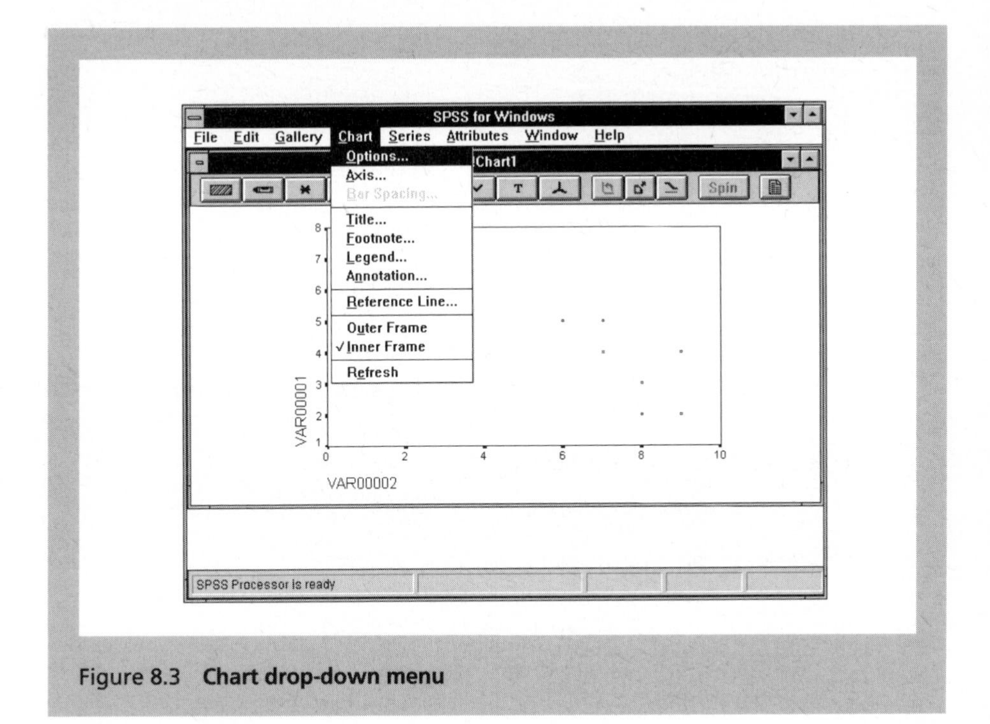

Figure 8.3 **Chart drop-down menu**

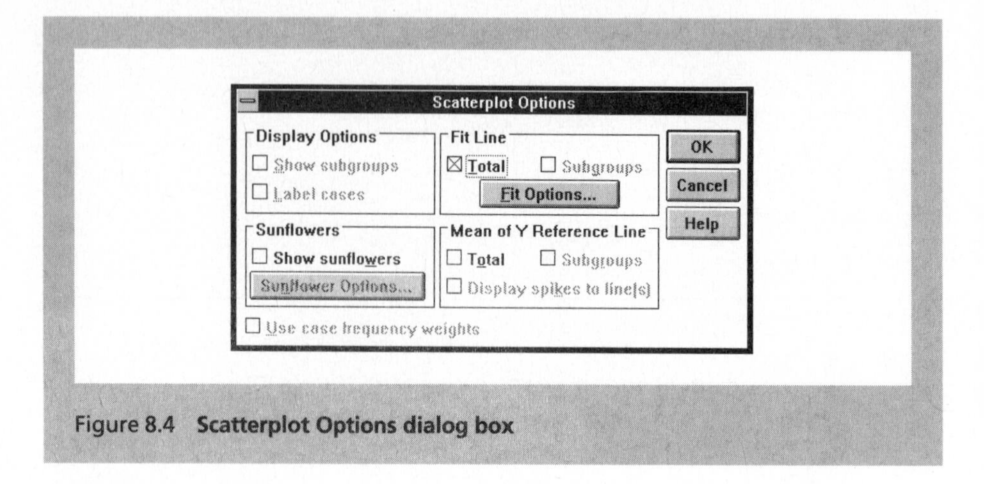

Figure 8.4 Scatterplot Options dialog box

- Select 'var00001' and the ▶ button next to the Y Axis: box which puts 'var00001' into this box as shown in Figure 7.3.
- Select 'var00002' and the ▶ button beside the X Axis: box which puts 'var00002' into this box as shown in Figure 7.3.
- Select OK which closes the Simple Scatterplot dialog box and the Newdata window and which displays the scatter diagram shown in Figure 7.4 in the Chart Carousel window.

To draw a regression line:

- Select Edit from the menu bar on the Chart Carousel window (Figure 2.11).
- Select Chart from the menu bar of the Chart Carousel window which reveals a drop-down menu (Figure 8.3).
- Select Options . . . from the Chart drop-down menu which opens the Scatterplot Options dialog box (Figure 8.4).
- Select Total in the Fit Line box.
- Select OK which closes the Scatterplot Options dialog box and produces the regression line in the scatterplot in the Chart Carousel window in Figure 8.5.

8.4 Interpreting Figure 8.5

This should be fairly obvious:

- The regression line sloping from the top left down to bottom right indicates a negative relationship between the two variables.
- The points seem relatively close to this line which suggests that the Beta weight (correlation) should be a large (negative) numerical value and that the confidence interval for the slope should be relatively small.

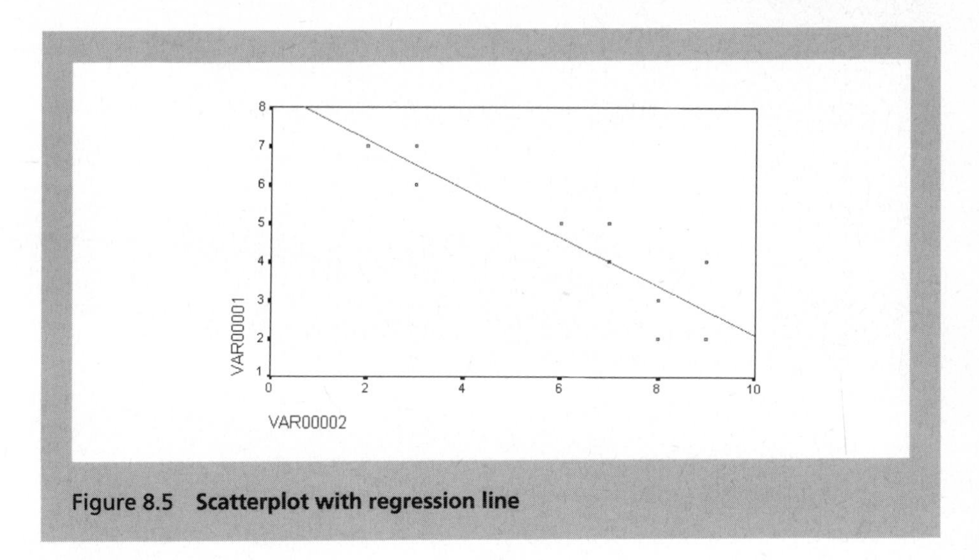

Figure 8.5 **Scatterplot with regression line**

8.5 Reporting Table 8.2 and Figure 8.5

Although all of the output from SPSS is pertinent to a sophisticated user, many readers might prefer to have just the bare bones at this stage.

- With this in mind, we would write about the analysis in this chapter: 'The scatterplot of the relationship between mathematical and musical ability suggested a linear negative relationship between the two variables. It is possible to predict accurately a person's musical ability from their mathematical ability. The equation is $Y' = 8.43 + (-0.63X)$ where X is an individual's mathematics score and Y' is the best prediction of their musical ability score.'
- An alternative is to give the scatterplot and to write underneath $a = 8.43$ and $B = -0.63$.

Chapter 9

Samples and populations

Generating a random sample

Random sampling is a key aspect of statistics. This chapter explains how random samples can be quickly generated.

In this chapter, the selection of random samples from a known set of scores is illustrated. The primary aim of this is to allow those learning statistics for the first time to try random sampling in order to get an understanding of sampling distributions. This, hopefully, will give a better understanding of estimation in statistics and the frailty that may underlie seemingly hard-nosed mathematical procedures. We will illustrate the generation of a random sample from a set of data consisting of the extraversion scores of the 50 airline pilots shown in Table 4.1.

9.1 Selecting a random sample

Quick summary

Data

Select Cases . . .

Random sample of cases

Sample . . .

Exactly

number of cases

total number of cases

Continue

OK

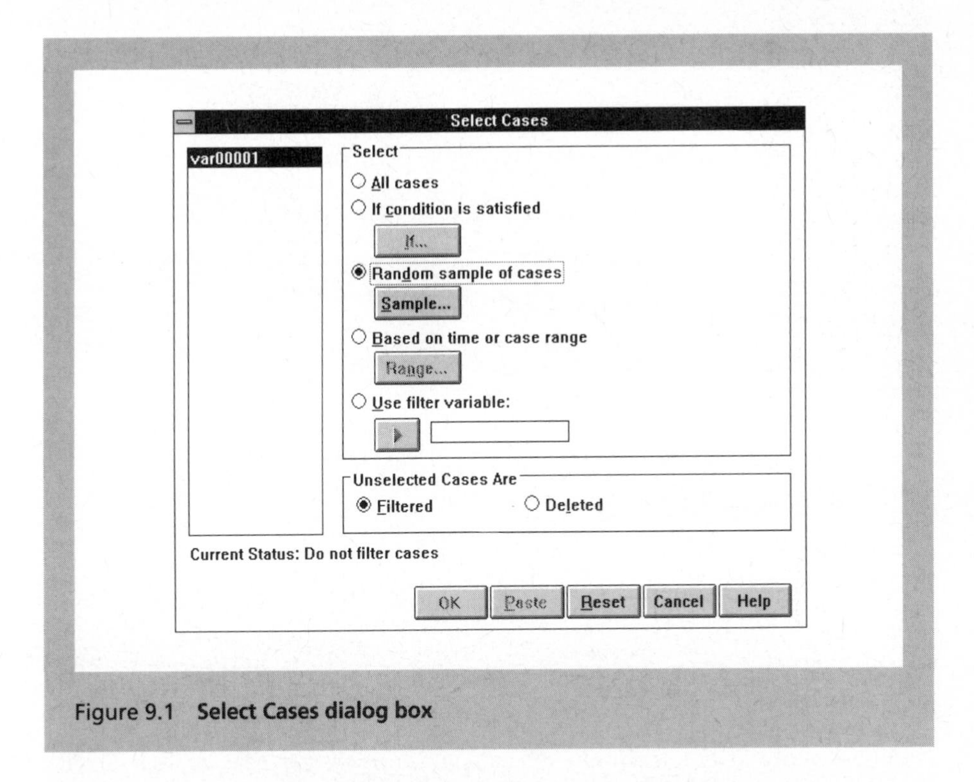

Figure 9.1 **Select Cases dialog box**

Figure 9.2 **Select Cases: Random Sample dialog box**

- As described in Chapter 1, type in the extraversion scores in the first column of Newdata or, if you have saved these data as a system file, call up this file.
- Select Data from the menu bar in the Applications window which produces a drop-down menu (Figure 2.2).
- Select Select Cases . . . from this drop-down menu which opens the Select Cases dialog box (Figure 9.1).

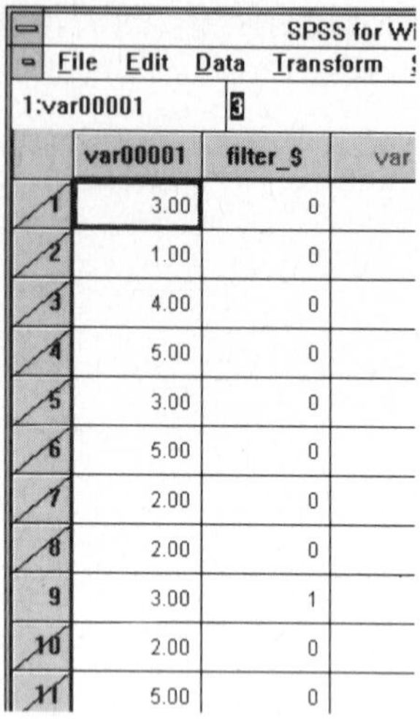

SPSS for Wi

File Edit Data Transform

1:var00001 3

	var00001	filter_$	var
1	3.00	0	
2	1.00	0	
3	4.00	0	
4	5.00	0	
5	3.00	0	
6	5.00	0	
7	2.00	0	
8	2.00	0	
9	3.00	1	
10	2.00	0	
11	5.00	0	

Figure 9.3 Random selection of cases

- Select Random sample of cases which opens the Select Cases: Random Sample dialog box (Figure 9.2).
- Select Sample . . .
- Select Exactly and for, say, a 10% sample of 50 cases, type 5 in the box beside it and 50 in the second box.
- Select Continue which closes the Select Cases: Random Sample dialog box.
- Select OK which closes the Select Cases dialog box and, after a few moments, opens the Newdata window (Figure 9.3).

9.2 Interpreting Figure 9.3

The second column is called filter_$ and consists of a series of 0's and 1's. The 1's represent the cases that have been selected and also do not have a line running through their row number, indicating that they have been selected.

9.3 Selecting the cases of the random sample

Quick summary

Data

Select Cases . . .

If condition is satisfied

filter_$

Continue

OK

- Select Data from the menu bar in the Applications window which produces a drop-down menu (Figure 2.2).
- Select Select Cases . . . from this drop-down menu which opens the Select Cases dialog box (Figure 9.1).
- Select If condition is satisfied which opens the Select Cases: If dialog box (Figure 9.4).
- Select 'filter_$' and then the ▶ button which puts 'filter_$' in the box beside it.
- Select Continue which closes the Select Cases: If dialog box.
- Select OK which closes the Select Cases dialog box and which opens the Output window. The random sample has now been selected.

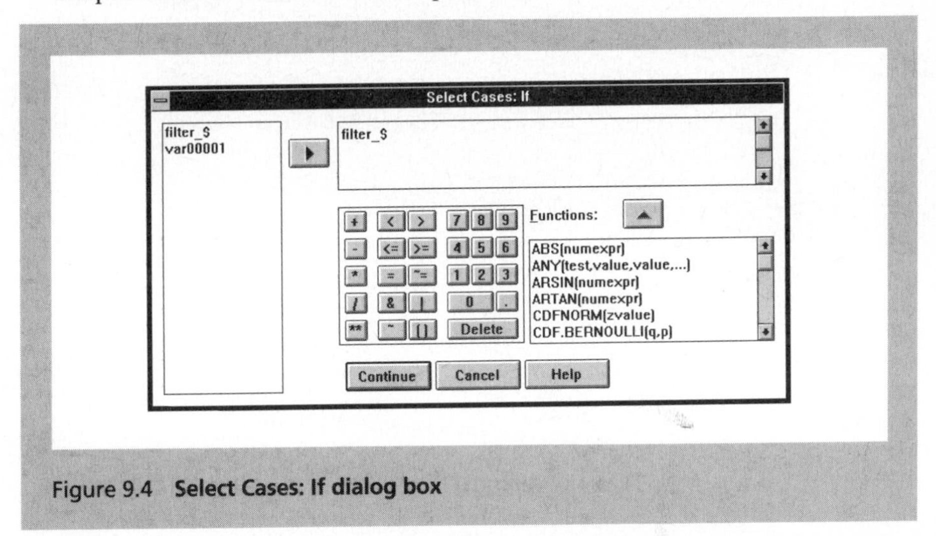

Figure 9.4 **Select Cases: If dialog box**

9.4 Statistical analysis on random sample

Quick summary

Statistics

Summarize

Descriptives . . .

var00001

▶

OK

- Select Statistics on the menu bar of the Applications window which produces a drop-down menu (Figure 2.6).
- Select Summarize from the Statistics drop-down menu which displays a second drop-down menu (Figure 2.6).
- Select Descriptives . . . which opens the Descriptives dialog box (Figure 5.1).
- Select 'var00001' and then the ▶ button which puts 'var00001' in the Variable[s]: text box.
- Select OK which closes the Descriptives dialog box and the Newdata window and which displays the output shown in Table 9.1 in the Output window.

Table 9.1 **Descriptive statistics produced by Descriptives**

Number of valid observations (listwise) = 5.00

Variable	Mean	Std Dev	Minimum	Maximum	Valid N	Label
VAR00001	3.40	1.52	1.00	5.00	5	

Chapter 10

Selecting cases

This chapter explains how to select a particular subgroup from your sample.

Sometimes we may wish to carry out computations on subgroups in our sample. For example we may want to correlate musical and mathematical ability (a) in girls and boys separately, (b) in older and younger children separately, and (c) in older and younger girls and boys separately. To do this, we need a code for sex and age such as 1 for girls and 2 for boys. We also need to decide what age we will use as a cut-off point to determine which children fall into the younger age group and which children fall into the older age group. We will use age 10 as the cut-off point with children aged 9 or less falling into the younger age group and children aged 10 or more falling into the older age group. Then we need to select each of the groups in turn and carry out the computation. We will illustrate the selection of cases with the data in Table 10.1 which shows the music and mathematics scores of 10 children together with their code for sex and their age in

Table 10.1 **Scores on musical ability and mathematical ability for 10 children with their sex and age**

Music score	Mathematics score	Sex	Age
2	8	1	10
6	3	1	9
4	9	2	12
5	7	1	8
7	2	2	11
7	3	2	13
2	9	2	7
3	8	1	10
5	6	2	9
4	7	1	11

years. (The music and mathematics scores are the same as those previously presented in Table 7.1.)

Obviously the selection of cut-off points is important. You need to beware of inadvertently excluding some cases.

10.1 Selecting cases

Quick summary

Data

Select Cases . . .

If condition is satisfied

If . . .

Select variable and ▶

Select condition (e.g. = 1)

Continue

OK

- Enter the data in Table 10.1 in Newdata, putting music scores in the first column, mathematics scores in the second column, the code for sex in the third column and the age in years in the fourth column as shown in Figure 10.1. If you saved the music and mathematics scores for the data in Chapter 7, you can retrieve this data file and add in the code for sex and the age in years.
- Select Data from the menu bar in the Applications window which produces a drop-down menu (Figure 2.2).
- Select Select Cases . . . from this drop-down menu which opens the Select Cases dialog box (Figure 9.1).
- Select If condition is satisfied.
- Select If . . . which opens the Select Cases: If dialog box (Figure 9.4).

- To select the girls, select VAR00003 and then the ▶ button which puts VAR00003 in the box beside it.
- Select = button which puts = sign after VAR00003.
- Select 1 which puts 1 after VAR00003 =.
- Select Continue which closes the Select Cases: If dialog box.

	var00001	var00002	var00003	var00004	var
1	2.00	8.00	1.00	10.00	
2	6.00	3.00	1.00	9.00	
3	4.00	9.00	2.00	12.00	
4	5.00	7.00	1.00	8.00	
5	7.00	2.00	2.00	11.00	
6	7.00	3.00	2.00	13.00	
7	2.00	9.00	2.00	7.00	
8	3.00	8.00	1.00	10.00	
9	5.00	6.00	2.00	9.00	
10	4.00	7.00	1.00	11.00	
11					

Figure 10.1 **Music and mathematics scores with sex and age in Newdata**

- Select OK which closes the Select Cases dialog box and returns to the Newdata window. The fifth column is called 'filter_$'. The rows or cases selected have a 1 in this fifth column. Cases not selected have a 0 in them. Girls have now been selected.
- Proceed with your statistical analysis (e.g. correlation).

- To select the boys next, select Data from the menu bar in the Applications window.
- Select Select Cases . . .
- Select If . . . var00003 = 1 which opens the Select Cases: If dialog box.
- Replace the 1 with a 2 so var00003 = 2.
- Select Continue which closes the Select Cases: If dialog box.
- Select OK which closes the Select Cases dialog box and returns to the Newdata window. Boys have now been selected.
- Proceed with your statistical analysis (e.g. correlation).

- To select the younger girls next, that is the girls of 9.0 years and younger, select Data from the menu bar in the Applications window.
- Select Select Cases . . .
- Select If . . . var00003 = 2 which opens the Select Cases: If dialog box.
- Replace the 2 with a 1 so var00003 = 1.
- Select &.

- Select var00004.
- Select <= (less than or equal to).
- Select 9.
- Select Continue which closes the Select Cases: If dialog box.
- Select OK which closes the Select Cases dialog box and returns to the Newdata window. Younger girls have now been selected.
- Proceed with your statistical analysis (e.g. correlation).

- To select the older girls next, that is the girls over 9.0 years, select Data from the menu bar in the Applications window.
- Select Select Cases . . .
- Select If . . . var00003 = 1 & var00004 <= 9 which opens the Select Cases: If dialog box.
- Replace <= with > (greater than).
- Select Continue which closes the Select Cases: If dialog box.
- Select OK which closes the Select Cases dialog box and returns to the Newdata window. Older girls have now been selected.
- Proceed with your statistical analysis (e.g. correlation).

Chapter 11

Standard error

Standard error is an index based on the standard deviation of the means of many samples taken from the population. In other words, it is a measure of the average amount by which the means of samples differ from the mean of the population from which they came. It is generally most used as an intermediate step in other statistical techniques, though it can be used like variance or standard deviation as an index of the amount of variability in the scores on a variable.

We will illustrate the computation of the estimated standard error of the mean with the set of six scores presented in Table 11.1 (*ISP* Table 11.3).

Table 11.1 **Data for standard error example**

X (scores)
5
7
3
6
4
5

11.1 Estimated standard error of the mean

Quick summary

Enter data

Statistics

Summarize

Descriptives . . .

var00001

▶

Options . . .

S.E.mean

Continue

OK

A number of SPSS procedures provide the standard error of the mean as part of the output.

- As described in Chapter 1, enter the six scores in Table 11.1 in the first column of the Newdata window.
- Select Statistics on the menu bar of the Applications window which produces a drop-down menu (Figure 2.6).
- Select Summarize from the Statistics drop-down menu which displays a second drop-down menu (Figure 2.6).
- Select Descriptives . . . which opens the Descriptives dialog box (Figure 5.1).
- Select 'var00001' and then the ▶ button which puts 'var00001' in the Variable[s]: text box.
- Select Options . . . which opens the Descriptives: Options sub-dialog box (Figure 5.2).
- Select S.E.mean.
- Select Continue which closes the Descriptives: Options sub-dialog box.
- Select OK which closes the Descriptives dialog box and the Newdata window and which displays the output shown in Table 11.2 in the Output window.

Table 11.2 **Standard error of the mean produced by Descriptives**

Number of valid observations (listwise) = 6.00

Variable	Mean	S.E. Mean	Std Dev	Minimum	Maximum	Valid N	Label
VAR00001	5.00	.58	1.41	3.00	7.00	6	

11.2 Interpreting the output in Table 11.2

- The table gives the value of the standard error of sample means as 0.58. This is the 'average' amount by which means of samples ($N = 6$) differ from the population mean.
- It is an estimate based on a sample and should really be termed the estimated standard error.
- The table includes other information such as the mean (5.00), the estimated population standard deviation based on this sample, and the minimum and maximum values in the data.

11.3 Reporting the output in Table 11.2

Generally, in psychological statistics, one would not report the standard error of sample means on its own. It would be more usual to report it as part of certain tests of significance. However, in many circumstances it is just as informative as the variance or standard deviation of a sample as it bears a simple relationship to both of these.

Chapter 12

The *t*-test

Comparing two samples of correlated/related scores

The correlated/related t-test tells you if the means of two sets of scores are significantly different from each other. It is used when the two sets of scores come from a single set of people or when the correlation coefficient between the two sets of scores is high.

We will illustrate the computation of a related *t*-test with the data in Table 12.1 which shows the number of eye-contacts made by the same babies with their mothers at 6 and 9 months (*ISP* Table 12.6).

Table 12.1 **Number of one-minute segments with eye-contact at different ages**

Subject	Six months	Nine months
Baby Clara	3	7
Baby Martin	5	6
Baby Sally	5	3
Baby Angie	4	8
Baby Trevor	3	5
Baby Sam	7	9
Baby Bobby	8	7
Baby Sid	7	9

12.1 Related *t*-test

Quick summary

Enter data

Statistics

Compare Means

Paired-Samples T Test . . .

var00001

var00002

OK

- As described in Chapter 1 enter the data of Table 12.1 into Newdata, putting the number of eye-contacts at 6 months in the first column and the number of eye-contacts at 9 months in the second column (Figure 12.1).
- Save this data set as it is used again in Chapter 18.
- Select Statistics on the menu bar of the Applications window which produces a drop-down menu (Figure 2.6).
- Select Compare Means from this drop-down menu which opens a second drop-down menu.
- Select Paired-Samples T Test . . . which opens the Paired-Samples T Test dialog box (Figure 12.2).
- Select 'var00001' which puts it beside Variable 1: in the Current Selections box.
- Select 'var00002' which puts it beside Variable 2: in the Current Selections box.

	var00001	var00002	var
1	3.00	7.00	
2	5.00	6.00	
3	5.00	3.00	
4	4.00	8.00	
5	3.00	5.00	
6	7.00	9.00	
7	8.00	7.00	
8	7.00	9.00	
9			

Figure 12.1 **Newdata containing amount of eye-contact at two ages**

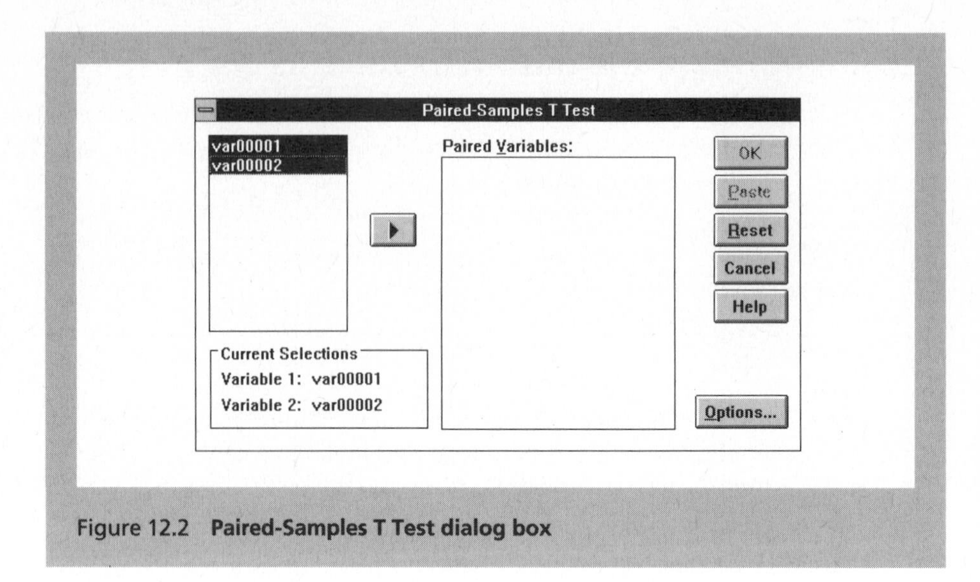

Figure 12.2 **Paired-Samples T Test dialog box**

- Select the ▶ button which puts 'var00001-var00002' in the Paired Variables: box.
- Select OK which closes the Paired-Samples T Test dialog box and the New-data window and which displays the output shown in Table 12.2 in the Output window.

Table 12.2 **Related *t*-test output**

t-tests for Paired Samples

Variable	Number of pairs	Corr	2-tail Sig	Mean	SD	SE of Mean
VAR00001				5.2500	1.909	.675
	8	.419	.301			
VAR00002				6.7500	2.053	.726

Paired Differences

Mean	SD	SE of Mean	t-value	df	2-tail Sig
–1.5000	2.138	.756	–1.98	7	.088
95% CI (–3.287, .287)					

12.2 Interpreting the output in Table 12.2

- In the upper part of the output the mean number of eye-contacts at 6 months ('var00001') and at 9 months ('var00002') is displayed under Mean. Thus the mean amount of eye contact is 5.2500 at 6 months and 6.7500 at 9 months. Rounded to two decimal places, the means are 5.25 and 6.75 respectively.
- In the upper part of the output is Corr which is the Pearson correlation coefficient between the two variables (eye-contact at 6 months and eye-contact at 9 months). *Ideally*, the value of this should be sizeable (in fact it is 0.419) and statistically significant (which it is not with a two-tailed significance level of 0.301). The related *t*-test assumes that the two variables are correlated and you might consider an unrelated *t*-test (Chapter 13) to be more suitable in this case.
- In the lower part of the output the difference between these two mean scores is presented under the Mean of Paired Differences and the standard error of this mean under SE of Mean. The difference between the two means is –1.5000 and the estimated standard error of means for this sample size is 0.756.
- The *t*-value of the difference between the sample means, its degrees of freedom and its two-tailed significance level are also shown in the lower part of the output. The *t*-value is –1.98 which has an exact two-tailed significance level of 0.088 with 7 degrees of freedom.

12.3 Reporting the output in Table 12.2

- We could report these results as follows: 'The mean number of eye-contacts at 6 months (M = 5.25, SD = 1.91) and that at 9 months (M = 6.75, SD = 2.05) did not differ significantly ($t = -1.98$, df = 7, two-tailed $p > 0.05$).'.
- Notice that the findings would have been statistically significant with a one-tailed test. However, it would have been necessary to predict this with sound reasons prior to being aware of the data. In this case one would have written to the effect that 'The two means differed significantly in the predicted direction ($t = -1.98$, df = 7, one-tailed $p < 0.05$).'.

Chapter 13

The *t*-test

Comparing two groups of unrelated/uncorrelated scores

The uncorrelated/unrelated* t*-test tells you if the means of two sets of scores are significantly different from each other. It is used when the two sets of scores come from two different samples of people.

We will illustrate the computation of an unrelated *t*-test with the data in Table 13.1 which shows the emotionality scores of 12 children from two-parent families and 10 children from single-parent families (*ISP* Table 13.8). In SPSS this sort of *t*-test is called an independent samples *t*-test.

Table 13.1 **Emotionality scores in two-parent and lone-parent families**

Two-parent family X_1	Lone-parent family X_2
12	6
18	9
14	4
10	13
19	14
8	9
15	8
11	12
10	11
13	9
15	
16	

13.1 Unrelated *t*-test

Quick summary

Enter data

Var00001 contains code for the two groups

Var00002 consists of the dependent variable

Statistics

Compare Means

Independent-Samples T Test . . .

Grouping variable (e.g. var00001)

Define Groups . . .

1

2

Continue

Dependent or test variable (e.g. var00002)

OK

- Take a good look at Figure 13.1. Notice that there are two columns of data. The second column ('var00002') consists of the 22 emotionality scores from *both* groups of children. The data are not kept separate for the two groups. In order to identify to which group the child belongs, the first column ('var00001') contains 1's and 2's. These indicate, in our example, children from lone-parent families (they are the rows with 1's in 'var00001') and children from two-parent families (they are the rows with 2's in 'var 00001'). Thus a single column is used for the dependent variable (in this case, emotionality, 'var00002') and another column for the independent variable (in this case, type of family, 'var00001'). So each row is for a particular child, and their independent variable and dependent variable scores are entered in two separate columns in Newdata.
- Save this data set as it is used again in Chapter 18.

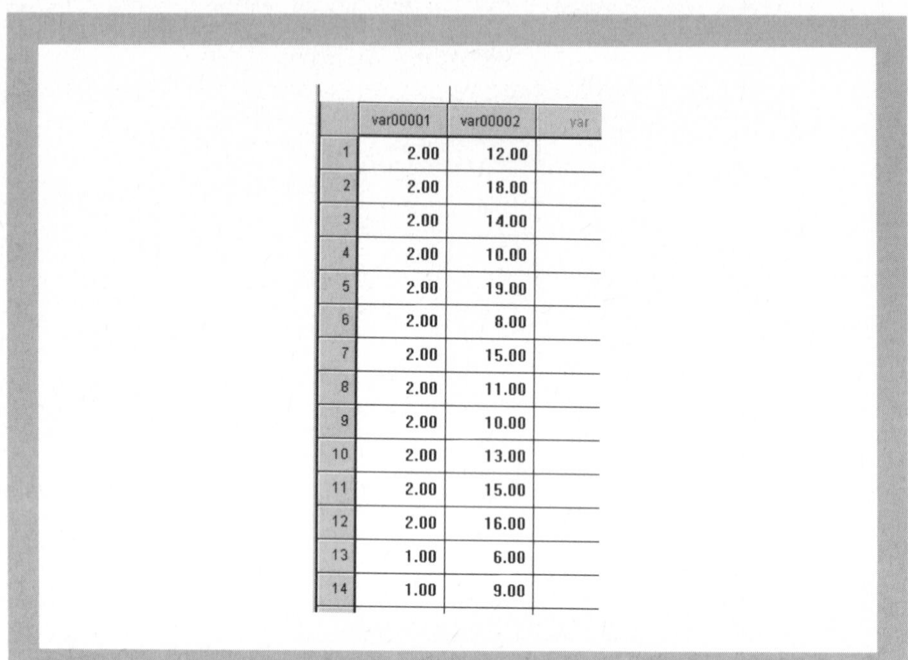

	var00001	var00002	var
1	2.00	12.00	
2	2.00	18.00	
3	2.00	14.00	
4	2.00	10.00	
5	2.00	19.00	
6	2.00	8.00	
7	2.00	15.00	
8	2.00	11.00	
9	2.00	10.00	
10	2.00	13.00	
11	2.00	15.00	
12	2.00	16.00	
13	1.00	6.00	
14	1.00	9.00	

Figure 13.1 **Newdata containing code for unrelated groups in var00001 and the scores of the dependent variable in var00002**

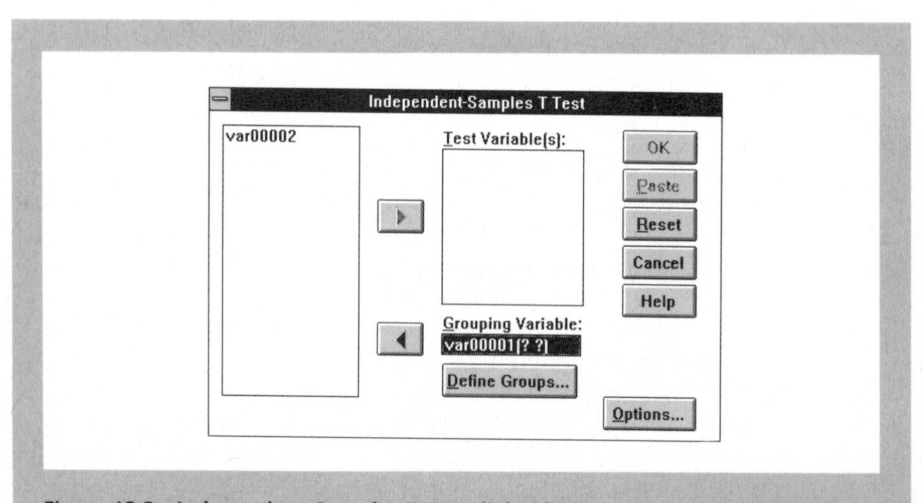

Figure 13.2 **Independent-Samples T Test dialog box**

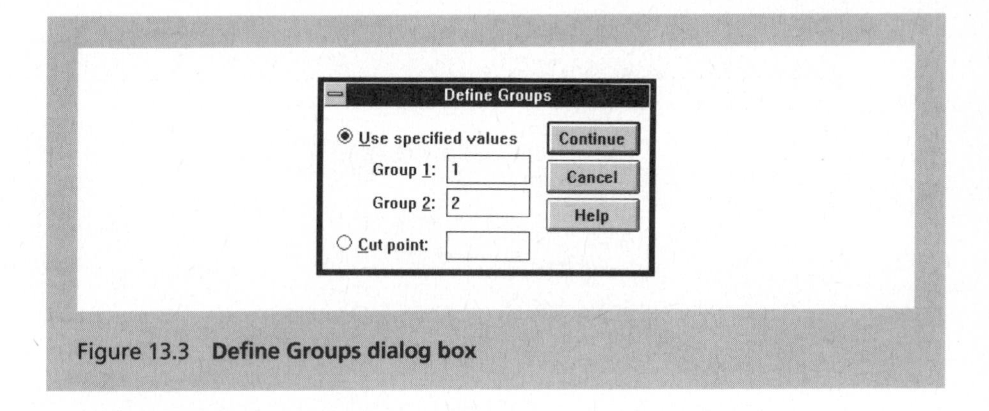

Figure 13.3 **Define Groups dialog box**

- Select Statistics on the menu bar of the Applications window which produces a drop-down menu (Figure 2.6).
- Select Compare Means from this drop-down menu which opens a second drop-down menu.
- Select Independent-Samples T Test . . . which opens the Independent-Samples T Test dialog box (Figure 13.2).
- Select 'var00001' and the ▶ button beside Grouping Variable: which puts 'var00001' in this box.
- Select Define Groups . . . which opens the Define Groups dialog box (Figure 13.3).
- Type 1 in the box beside Group 1:.
- Select box beside Group 2: and type 2.
- Select Continue which closes the Define Groups dialog box.
- Select 'var00002' and the ▶ button beside Test Variable[s]: which puts 'var00002' in this box.
- Select OK which closes the Independent-Samples T Test dialog box and the Newdata window and which displays the output shown in Table 13.2 (overleaf).

13.2 Interpreting the output in Table 13.2

The output for the uncorrelated/unrelated *t*-test on SPSS is particularly confusing even to people with a good knowledge of statistics. The reason is that there are two versions of the uncorrelated/unrelated *t*-test. Which one to use depends on whether or not there is a significant difference between the (estimated) variances for the two groups of scores.

- Examine the upper part of the output. This contains the means and standard deviations of the scores on the dependent variable (emotionality) of the two groups. Notice that an additional figure has been added by the computer to

Table 13.2 **Unrelated *t*-test output**

```
t-tests for Independent Samples of VAR00001

                Number
Variable        of Cases          Mean          SD      SE of Mean

VAR00002

VAR00001 1          10          9.5000       3.100          .980
VAR00001 2          12         13.4167       3.370          .973

     Mean Difference = –3.9167
     Levene's Test for Equality of Variances: F= .212     P= .650

     t-test for Equality of Means
                                                              95%
Variances  t-value      df  2-Tail Sig  SE of Diff      CI for Diff

Equal        –2.81  20            .011      1.392  (–6.821, –1.013)
Unequal      –2.84  19.77         .010      1.381  (–6.800, –1.034)
```

the name of the column containing the dependent variable. This additional figure indicates which of the two groups the row refers to.

- For children from two-parent families ('VAR00001 2') the mean emotionality score is 13.4167 and the standard deviation of the emotionality scores is 3.370. Rounded to two decimal places, these values are 13.42 and 3.37. For the children of lone-parent families ('VAR00001 1') the mean emotionality score is 9.5000 and the standard deviation of emotionality is 3.100. Rounded to two decimal places these values are 9.50 and 3.10.
- Read the line referring to Levene's Test for Equality of Variances. If the probability value is statistically significant then your variances are *unequal*. Otherwise they are equal.
- Levene's test for equality of variances in this case tells us that the variances are equal because the value P = 0.650 is not statistically significant.
- Finally look at the last section of the output. You need the row for Equal variances and you can delete the row for Unequal variances. The *t* value, its degrees of freedom and its probability are displayed in the lower part of the output. The *t* value for equal variances is –2.81, which with 20 degrees of freedom has an exact two-tailed significance level of 0.011.
- Had Levene's test for equality of variances been statistically significant (i.e. 0.05 or less), then you should have used the final row of the output which gives the *t*-test values for unequal variances.

13.3 Reporting the output in Table 13.2

- We could report the results of this analysis as follows: 'The mean emotionality score of children from two-parent families (M = 13.42, SD = 3.37) is significantly higher ($t = -2.81$, df = 20, two-tailed $p<0.05$) than that of children in lone-parent families (M = 9.50, SD = 3.10).'.
- It is unusual to see the *t*-test for unequal variances in psychological reports. Many psychologists are unaware of its existence. So what happens if you have to use one? In order to clarify things, we would write: 'Because the variances for the two groups were significantly unequal ($F = 8.43$, $p< 0.05$), a *t*-test for unequal variances was used . . .'.

Chapter 14

Chi-square

Differences between samples of frequency data

Chi-square tells you if two or more samples each consisting of frequency data (nominal data) differ from each other. It can also be used to test whether a single sample differs significantly from a known population.

We will illustrate the computation of chi-square with two or more samples with the data in Table 14.1 (*ISP* Table 14.8). This table shows which one of three types of television programme is favoured by a sample of 119 teenage boys and girls. To analyse a table of data like this with SPSS we first have to input the data into Newdata and weight the cells by the frequencies of cases in them.

If we are working with a ready-made table, it is necessary to go through the 'Weighting data' procedure first (Figure 14.1). Otherwise, you would enter the above table case by case, indicating which category of the row and which category of the column each case belongs to (see Figure 14.2).

Table 14.1 **Relationship between favourite type of TV programme and sex of respondent**

Respondents	Soap opera	Crime drama	Neither
Males	27	14	19
Females	17	33	9

14.1 Weighting data

(Ignore this section if you are not using a ready-made table.)

	var00001	var00002	var00003	var
1	1.00	1.00	27.00	
2	1.00	2.00	14.00	
3	1.00	3.00	19.00	
4	2.00	1.00	17.00	
5	2.00	2.00	33.00	
6	2.00	3.00	9.00	
7				

Figure 14.1 Weighted cases in Newdata

Quick summary

Enter code for rows in first column

Enter code for columns in second column

Enter frequencies in third column

Data

Weight Cases . . .

Weight cases by

var00003

OK

- We need to identify each of the six cells in Table 14.1. The rows of the table represent the sex of the participants while the columns represent the three types of television programme. We will then weight each of the six cells of the table by the number of cases in them.
- The first column contains the code for males (1) and females (2) as shown in Figure 14.1.
- The second column holds the code for the three types of television programme: soap opera (1); crime drama (2); and neither (3).
- The third column has the number of people in each of these six cells.

	sex	tvprog	var
1	1.00	1.00	
2	1.00	2.00	
3	1.00	3.00	
4	2.00	1.00	
5			

Figure 14.2 **Unweighted cases in Newdata**

- Select Data from the menu bar in the Applications window which produces a drop-down menu (Figure 2.2).
- Select Weight Cases . . . from this drop-down menu which opens the Weight Cases dialog box (Figure 2.3).
- Select Weight cases by.
- Select 'var00003' and then the ▶ button which puts 'var00003' in the Frequency Variable: text box.
- Select OK which closes the Weight Cases dialog box. The six cells are now weighted by the numbers in the third column.

14.2 Chi-square for two or more samples

In this chapter we have concentrated on how one can analyse data from pre-existing contingency tables. This is why we need the weighting procedure. However, you will not always be using ready-made tables. Any variables which consist of just a small number of nominal categories can be used for chi-square. For example, if one wished to examine the relationship between sex (coded 1 for male, 2 for female) and age (coded 1 for under 20 years, 2 for 20 to 39 years, and 3 for 40 years and over), the procedure is as follows: (a) enter the age codes for your, say, 60 cases in the first column of Newdata; (b) enter the age categories for each of these cases in the equivalent row of the next column. You can then carry out your chi-square as follows. You do not go through the weighting procedure first. The frequencies in the cells are calculated by the computer for you.

Quick summary

Statistics

Summarize

Crosstabs . . .

var00001

▶ *Row[s]:*

var00002

▶ *Column[s]:*

Statistics . . .

Chi-square

Continue

Cells . . .

Expected

Continue

OK

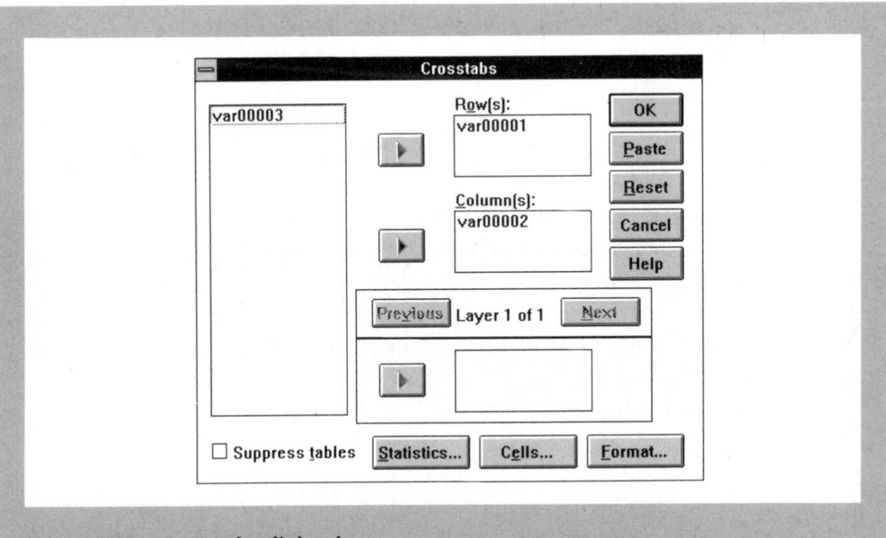

Figure 14.3 **Crosstabs dialog box**

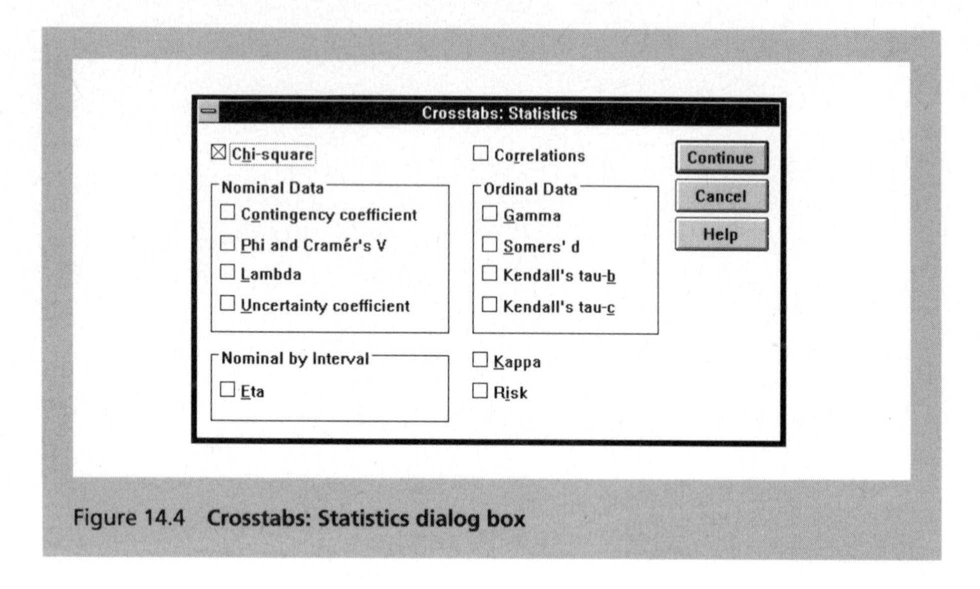

Figure 14.4 **Crosstabs: Statistics dialog box**

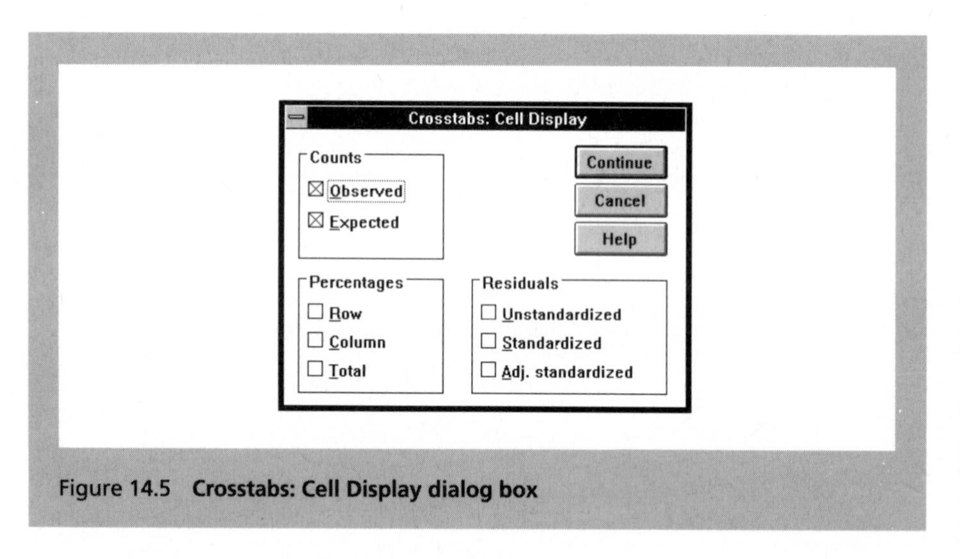

Figure 14.5 **Crosstabs: Cell Display dialog box**

- Select Statistics on the menu bar of the Applications window which produces a drop-down menu (Figure 2.6).
- Select Summarize from the Statistics drop-down menu which displays a second drop-down menu (Figure 2.6).
- Select Crosstabs . . . which opens the Crosstabs dialog box (Figure 14.3).
- Select 'var00001' and then the ▶ button beside Row[s]: which puts 'var00001' in this box.

- Select ‘var00002’ and then the ▶ button beside Column[s]: which puts ‘var00002’ in this box.
- Select Statistics . . . which opens the Crosstabs: Statistics dialog box (Figure 14.4).
- Select Chi-square.
- Select Continue which closes the Crosstabs: Statistics dialog box.
- Select Cells . . . which opens the Crosstabs: Cell Display dialog box (Figure 14.5).
- Select Expected in the Counts box.
- Select Continue which closes the Crosstabs: Cell Display dialog box.
- Select OK which closes the Crosstabs dialog box and the Newdata window and which displays the output shown in Table 14.2 in the Output window.

Table 14.2 **Chi-square output produced by Crosstabs**

VAR00001 by VAR00002

VAR00002 Page 1 of 1

VAR00001 (Count Exp Val)	1.00	2.00	3.00	Row Total
1.00	27 22.2	14 23.7	19 14.1	60 50.4%
2.00	17 21.8	33 23.3	9 13.9	59 49.6%
Column Total	44 37.0%	47 39.5%	28 23.5%	119 100.0%

Chi-Square	Value	DF	Significance
Pearson	13.51756	2	.00116
Likelihood Ratio	13.84088	2	.00099
Mantel-Haenszel test for linear association	.00026	1	.98722

Minimum Expected Frequency – 13.882

Number of Missing Observations: 0

14.3 Interpreting the output in Table 14.2

- The table shows the observed and expected frequency of cases in each cell. The observed frequency (called Count) is presented first and the expected frequency (called Exp Val) second. The observed frequencies are always whole numbers so they should be easy to spot. The expected frequencies are always expressed to one decimal place so they are easily identified. Thus the first cell of the table (defined by 1.00 for 'var00001' and for 'var00002') has an observed frequency of 27 and an expected frequency of 22.2.
- The Row Total lists the number of cases in that row followed by this number expressed as a percentage of the total number of cases in the table. So the first row has 60 cases which expressed as a percentage of the total number of cases in the table (119 cases) is 50.4%.
- The Column Total first presents the number of cases in that column followed by this number expressed as a percentage of the total number of cases in the table for that column. Thus the first column has 44 cases which represents 37.0 % of the number of cases in the table (119 cases).
- **The chi-square value, its degrees of freedom and its significance level are displayed on the line starting with the word Pearson, the man who developed this test. The chi-square value is 13.51756 which, rounded to two decimal places, is 13.52. Its degrees of freedom are 2 and its exact two-tailed probability is 0.00116.**
- Also shown in the output is the Minimum Expected Frequency of any cell in the table which is 13.882 for the last cell (defined by 2.00 for 'var00001' and 3.00 for 'var00002'). If the minimum expected frequency is less than 5.0 then we should be wary of using chi-square. If you have a 2 × 2 chi-square and small expected frequencies occur, it would be better to use the Fisher Exact Test (see below) which SPSS prints in the output in these circumstances.

14.4 Reporting the output in Table 14.2

There are two alternative ways of describing these results. To the inexperienced eye they may seem very different but they amount to the same thing:

- We could describe the results in the following way: 'There was a significant difference between the observed and expected frequency of teenage boys and girls in their preference for the three types of television programme ($\chi^2 = 13.52$, df = 2, $p<0.01$).'.
- Alternatively, and just as accurate: 'There was a significant association between sex and preference for different types of television programme ($\chi^2 = 13.52$, df = 2, $p<0.01$).'.

14.5 Fisher's Exact Test

The chi-square procedure computes Fisher's Exact Test for 2 × 2 tables when one or more of the four cells have an expected frequency of less than 5. Fisher's Exact Test would be computed for the data in Table 14.3 (*ISP* Table 14.14). The SPSS output for this table is presented in Table 14.4.

Table 14.3 **Photographic memory and sex of respondents**

Respondents	Photographic memory	No photographic memory
Males	2	7
Females	4	1

Table 14.4 **Fisher's Exact Test probabilities produced by Crosstabs for 2 × 2 table with expected frequencies of less than 5**

```
VAR00001 by VAR00002
                    VAR00002                 Page 1 of 1
           Count  |
         Exp Val  |
                  |                  Row
                  |  1.00  |  2.00  | Total
VAR00001   -------+--------+--------+
            1.00  |    2   |    7   |    9
                  |   3.9  |   5.1  | 64.3%
                  +--------+--------+
            2.00  |    4   |    1   |    5
                  |   2.1  |   2.9  | 35.7%
                  +--------+--------+
          Column       6        8       14
           Total   42.9%    57.1%   100.0%

    Chi-Square                  Value        DF      Significance
Pearson                        4.38148        1         .03633
Continuity Correction          2.33981        1         .12610
Likelihood Ratio               4.58269        1         .03230
Mantel-Haenszel test for       4.06852        1         .04369
    linear association
Fisher's Exact Test:
  One-Tail                                              .06294
  Two-Tail                                              .09091

Minimum Expected Frequency — 2.143
Cells with Expected Frequency <5 —    3 OF    4 (75.0%)

Number of Missing Observations: 0
```

14.6 Interpreting the output in Table 14.4

The significance of Fisher's Exact Test for this table is 0.06294 at the one-tailed level and 0.09091 at the two-tailed level.

14.7 Reporting the output in Table 14.4

- We would write: 'There was no significant relationship between sex and the possession of a photographic memory (Fisher exact probability = 0.09, two-tailed test).' or 'Males and females do not differ in the frequency of possession of a photographic memory (Fisher exact probability = 0.09, two-tailed test).'.
- However, with such a small sample size, the finding might best be regarded as marginally significant and a strong recommendation made that further studies should be carried out in order to establish with more certainty whether girls actually do possess photographic memories more frequently.

14.8 One-sample chi-square

Quick summary

Enter code for categories in first column of Newdata

Enter frequency of cases in second column

Data

Weight Cases . . .

Weight cases by

var00001

OK

Statistics

Nonparametric Tests

Chi-square . . .

var00001

Values:

Type expected value

Add

Repeat as needed

OK

Table 14.5 **Data for a one-sample chi-square**

	Clear smilers	Clear non-smilers	Impossible to classify
Observed frequency	35	40	5
Expected frequency	40	32	8

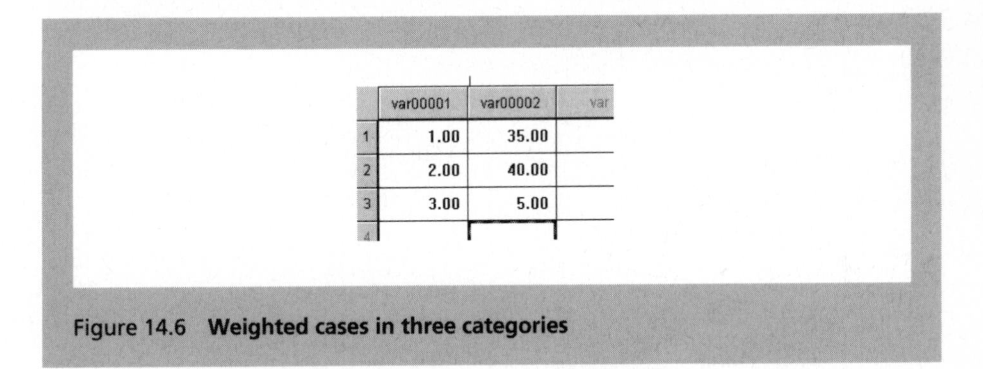

Figure 14.6 **Weighted cases in three categories**

We will illustrate the computation of a one-sample chi-square with the data in Table 14.5 (*ISP* Table 14.16) which shows the observed and expected frequency of smiling in 80 babies. The expected frequencies were obtained from an earlier large-scale study.

- In the first column of Newdata enter the code for the three categories of smilers as shown in Figure 14.6.
- In the second column enter the observed frequency of babies falling in the three categories.
- Weight the three categories as described at the beginning of this chapter.
- Select Statistics on the menu bar of the Applications window which produces a drop-down menu (Figure 2.6).
- Select Nonparametric Tests from this drop-down menu which opens a second drop-down menu.
- Select Chi-square . . . which opens the Chi-Square Test dialog box (Figure 14.7).
- Select 'var00001' and then the ▶ button which puts 'var00001' in the Test Variable List: box.
- Select Values: in the Expected Values section.
- Type 40 in the Values: box.
- Select Add.

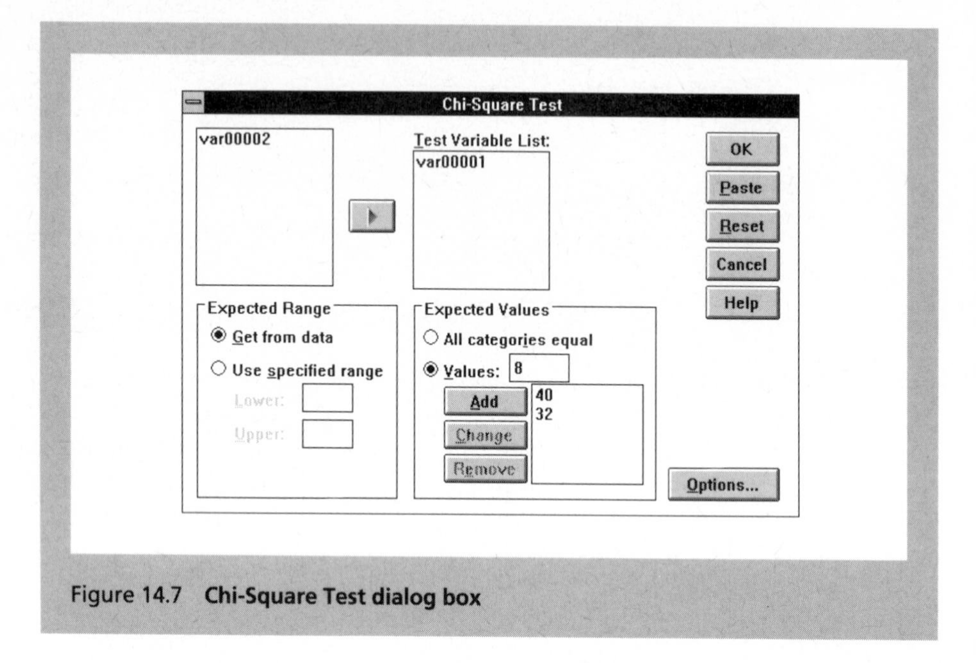

Figure 14.7 **Chi-Square Test dialog box**

- Type 32 in the Values: box.
- Select Add.
- Type 8 in the Values: box.
- Select Add.
- Select OK which closes the Chi-Square Test dialog box and the Newdata window and which displays the output shown in Table 14.6 in the Output window.

Table 14.6 **One-sample chi-square output produced by Chi-Square**

```
-----Chi-Square Test

VAR00001

                Cases
Category      Observed        Expected        Residual
    1.00            35           40.00           -5.00
    2.00            40           32.00            8.00
    3.00             5            8.00           -3.00
                   ---
   Total            80

Chi-Square                D.F.                Significance
    3.7500                   2                       .1534
```

14.9 Interpreting the output in Table 14.6

- The codes of the three categories are shown in the first column under the heading Category.
- The observed frequencies of cases are presented in the second column under the heading Cases Observed.
- The expected frequencies of cases are displayed in the third column under the heading Expected.
- The differences or residuals between the observed and expected frequencies are listed in the fourth column under the heading Residual.
- The value of chi-square, its degrees of freedom and its significance are presented in the last line of the output. Chi-square, rounded to two decimal places, is 3.75, its degrees of freedom are 2 and its exact significance level is 0.1534.

14.10 Reporting the output in Table 14.6

We could describe the results of this analysis as follows: 'There was no statistical difference between the observed and expected frequencies for the three categories of smiling in infants ($\chi^2 = 3.75$, df = 2, ns).'.

14.11 McNemar test

We will illustrate the computation of McNemar's test with the data in Table 14.7 which shows the number of teenage students who changed or did not change their minds about going to university after listening to a careers talk favouring university education (*ISP* Table 14.17). The table gives the numbers who wanted to go to university before the talk and after it (30), those who wanted to go before the talk but not after it (10), those who wanted to go to university after the talk but not before it (50), and the numbers not wanting to go to university both before and after the talk (32).

Table 14.7 **Students wanting to go to university before and after a careers talk**

	Before talk 'yes'	Before talk 'no'
After talk 'yes'	30	50
After talk 'no'	10	32

Quick summary

Enter code for rows in first column

Enter code for columns in second column

Enter frequencies in third column

Data

Weight Cases . . .

Weight cases by

var00003

OK

Statistics

Nonparametric Tests

2 Related Samples . . .

var00001

var00002

Wilcoxon

McNemar

OK

- In the first column of Newdata enter the code for the two rows representing the 'after the talk' position as shown in Figure 14.8.
- In the second column enter the code for the two columns representing the 'before the talk' position.
- In the third column enter the observed frequency of teenagers falling into the four cells of the table.
- Weight the four cells as described at the beginning of this chapter.
- Select Statistics on the menu bar of the Applications window which produces a drop-down menu (Figure 2.6).
- Select Nonparametric Tests from this drop-down menu which opens a second drop-down menu.

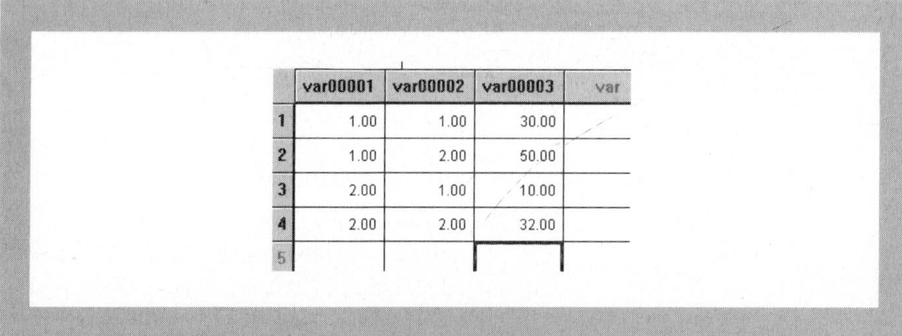

	var00001	var00002	var00003	var
1	1.00	1.00	30.00	
2	1.00	2.00	50.00	
3	2.00	1.00	10.00	
4	2.00	2.00	32.00	
5				

Figure 14.8 **Weighted cells for a McNemar test**

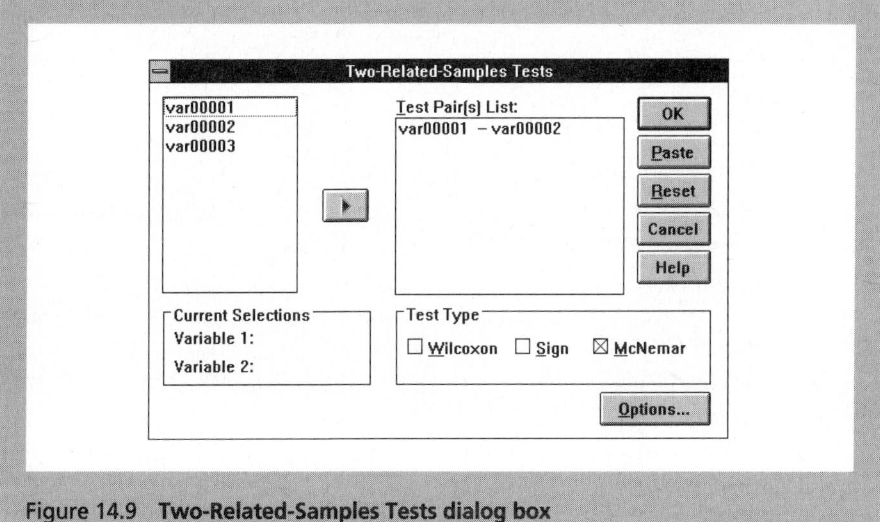

Figure 14.9 **Two-Related-Samples Tests dialog box**

- Select 2 Related Samples . . . which opens the Two-Related-Samples Tests dialog box (Figure 14.9).
- Select 'var00001' which puts it next to Variable 1: in the Current Selections section.
- Select 'var00002' which puts it next to Variable 2: in the Current Selections section.
- Select the ▶ button which puts 'var00001–var00002' in the Test Pair[s] List: box.
- Select Wilcoxon in the Test Type section to de-activate it.
- Select McNemar.
- Select OK which closes the Two-Related-Samples Tests dialog box and the Newdata window and which displays the output shown in Table 14.8 in the Output window.

Table 14.8 **McNemar test output produced by 2 Related Samples**

```
-----McNemar Test

     VAR00001
with VAR00002

                              VAR00002
                           2.00      1.00        Cases              122
                        +-------+-------+
                1.00    |   50  |   30  |  Chi-Square          25.3500
VAR00001                +-------+-------+
                2.00    |   32  |   10  |  Significance          .0000
                        +-------+-------+
```

14.12 Interpreting the output in Table 14.8

- In the first cell of the table (defined by 1.00 for 'var00001' and 2.00 for 'var00002') is the number of teenagers who changed from not wanting to go to university before hearing the talk to wanting to go to university after hearing the talk.
- In the last cell of the table (defined by 2.00 for 'var00001' and 1.00 for 'var00002') is the number of teenagers who changed from wanting to go to university before hearing the talk to not wanting to go to university after hearing the talk.
- The chi-square value, rounded to two decimal places, is 25.35 and its significance level is less than 0.00005.

14.13 Reporting the output in Table 14.8

We can report the results of this analysis as follows: 'There was a significant increase in the number of teenagers who wanted to go to university after hearing the talk ($\chi^2 = 25.35$, df = 1, $p<0.0001$).'.

Chapter 15

Missing values

Sometimes in research, you may not have a complete set of data from each participant. Missing values tells the computer how to deal with such situations.

15.1 Introduction

When collecting data, information for some of the cases on some of the variables is often missing. Take, for example, the data in Table 15.1 which consist of the music and mathematics scores of 10 children with their code for gender and their age in years. There is no missing information for any of the four variables for any of the 10 cases. But suppose that the first two cases were away for the music test so that we had no scores for them. It would be a pity to discard all the data for these two cases given that we have information on them for the other three variables of mathematics, gender and age. Consequently we would enter the available data for these other variables. Although we could leave the music score cells empty for these two cases, what we usually do is to

Table 15.1 **Scores on musical ability and mathematical ability for 10 children with their sex and age**

Music score	Mathematics score	Sex	Age
2	8	1	10
6	3	1	9
4	9	2	12
5	7	1	8
7	2	2	11
7	3	2	13
2	9	2	7
3	8	1	10
5	6	2	9
4	7	1	11

code missing data with a number which does not correspond to any possible value that the variable could take.

Suppose the scores for the music test can vary from 0 to 10. We can use any number, other than 0 to 10, to signify a missing value for the music test. We will use the number 11 as the code for a missing music score so that the values in the first two rows of the first column are 11 in Newdata. We will also assume that the age for the third case is missing. We will use the number 0 as the code for age which is missing. Now we need to tell SPSS how we have coded missing data. If we do not do this, then SPSS will read these codes as real numbers.

Missing values can also be used to tell the computer to ignore certain values of a variable which you wish to exclude from your analysis. So, for example, you could use missing values in relation to chi-square to get certain categories ignored.

Leaving a cell blank in the Newdata spreadsheet results in a full-stop (.) being entered in the cell if it is part of the active matrix of entries. On the output these are identified as missing values but they are best regarded as omitted values. We would recommend that you do not use blank cells as a way of identifying missing values since it does not distinguish between truly missing values and keyboard errors. Normally, substantial numbers such as 99 or 999 are the best way of identifying a missing value.

15.2 Defining missing values

Quick summary

Variable column

Data

Define Variable . . .

Missing Values . . .

Discrete missing values

Code

Continue

OK

- Enter the data in Table 15.1 in Newdata, including the missing values (11 and 0 respectively) for the music test and age (Figure 15.1). If you saved the data used in Chapter 10, then you can retrieve these data.

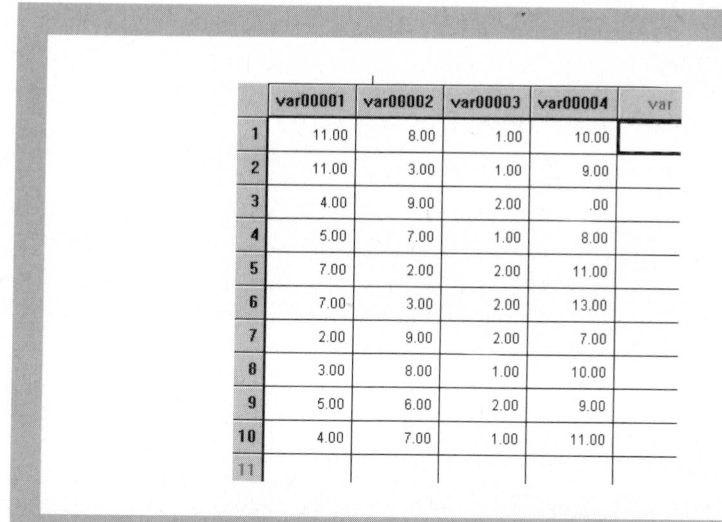

	var00001	var00002	var00003	var00004	var
1	11.00	8.00	1.00	10.00	
2	11.00	3.00	1.00	9.00	
3	4.00	9.00	2.00	.00	
4	5.00	7.00	1.00	8.00	
5	7.00	2.00	2.00	11.00	
6	7.00	3.00	2.00	13.00	
7	2.00	9.00	2.00	7.00	
8	3.00	8.00	1.00	10.00	
9	5.00	6.00	2.00	9.00	
10	4.00	7.00	1.00	11.00	
11					

Figure 15.1 **Music and mathematics scores with gender and age in Newdata with missing data for music scores and age**

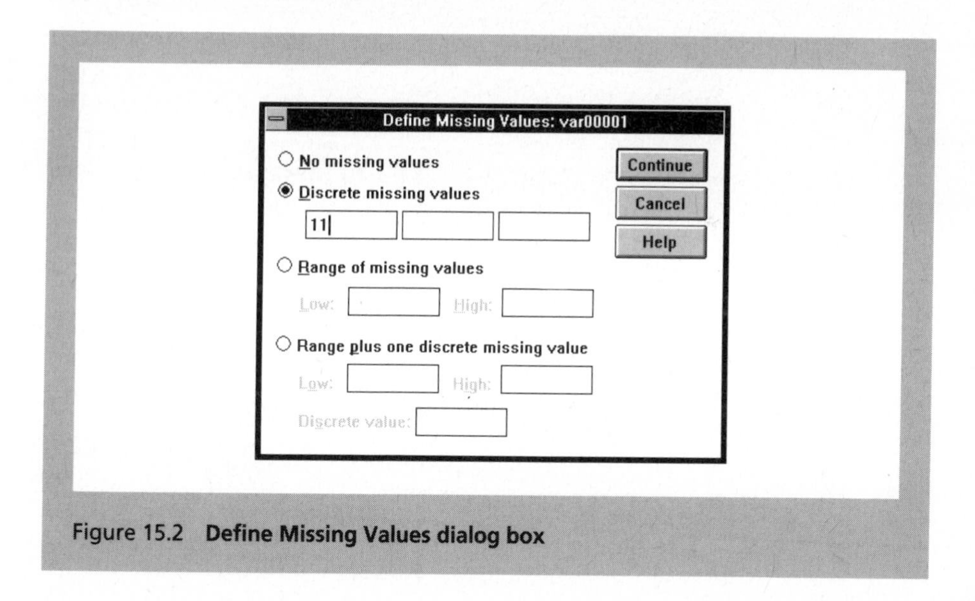

Figure 15.2 **Define Missing Values dialog box**

- To code missing values for a variable, put the cursor anywhere in the Newdata column which holds the values of that variable. So, to code missing values for the music test put the cursor in the first column.
- Select Data from the menu bar in the Applications window which produces a drop-down menu (Figure 2.2).

- Select Define Variable . . . which opens the Define Variable dialog box (Figure 2.4).
- Select Missing Values . . . in the Change Settings section which opens the Define Missing Values dialog box (Figure 15.2).
- Select Discrete missing values and type 11 in the first box.
- Select Continue which closes the Define Missing Values dialog box and puts 11 beside Missing Values in the Variable Description section of the Define Variable dialog box.
- Select OK which closes the Define Variable dialog box and opens the New-data window.
- Repeat the procedure for the fourth column, typing 0 in the first box.

15.3 Pairwise and listwise options

We will illustrate some of the options available when you have missing data with the Correlate procedure, although similar kinds of options are available with some of the other statistical procedures.

- Select Statistics which produces a drop-down menu (Figure 2.6).
- Select Correlate from the drop-down menu which reveals a smaller drop-down menu.
- Select Bivariate from this drop-down menu which opens the Bivariate Correlations dialog box (Figure 7.1).
- Select 'var00001', 'var00002', 'var00003' and 'var00004' and then the ▶ button which puts these variables in the Variables: text box.
- Select Options . . . which opens the Bivariate Correlations: Options dialog box (Figure 15.3).

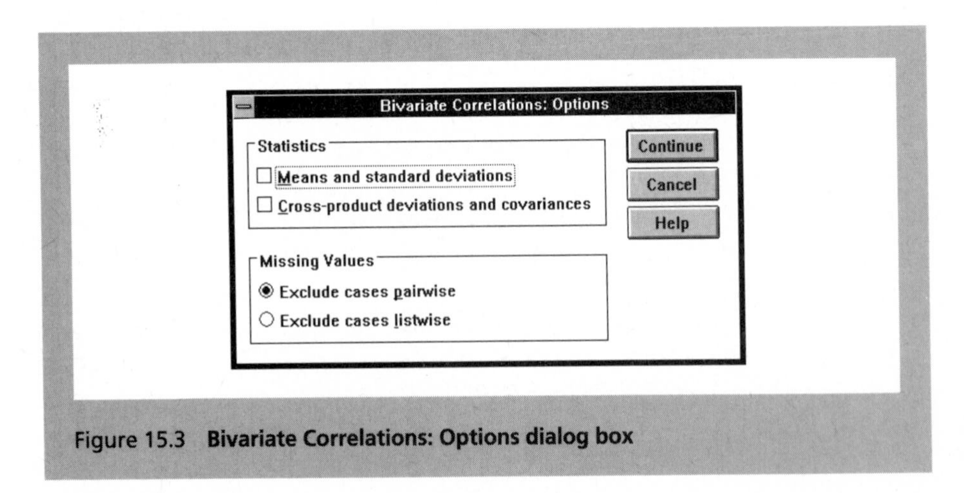

Figure 15.3 **Bivariate Correlations: Options dialog box**

- The default missing values option is Exclude cases pairwise. This means that a correlation will be computed for all cases which have non-missing values for any two or a pair of variables. Since there are two missing values for the music test and no missing values for the mathematics test and gender, the number of cases on which these correlations involving music scores will be based is 8 as shown in Table 15.2. Since one value for age is missing for another case, the number of cases on which the correlation between music scores and age is based is 7. As there are no missing values for the mathematics test and gender, the number of cases on which this correlation is based is 10. The number of cases on which the correlation between the mathematics score and age is based is 9 since there is one missing value for age and none for mathematics. Finally, although we are not really concerned about the correlation between gender and age, the number of cases here is also 9 since there is one missing value for age and none for gender.
- *Notice that the number of cases varies for pairwise deletion of missing values.*
- The alternative missing values option is Exclude cases listwise in which correlations are computed for all cases which have no missing values on any of the variables which have been selected for this procedure. In this example, the number of cases which have no missing values on any of the four variables selected is 7 as shown in Table 15.3 which presents the output for this option.
- *Notice that the number of cases does not vary for listwise deletion of missing values.*

Table 15.2 **Exclude cases pairwise output from Correlate**

```
-- Correlation Coefficients --

              VAR00001     VAR00002     VAR00003     VAR00004

VAR00001       1.0000       -.9232        .2928        .6809
              (    8)      (    8)      (    8)      (    7)
              P= .         P= .001      P= .482      P= .092

VAR00002       -.9232       1.0000       -.1612       -.5501
              (    8)      (   10)      (   10)      (    9)
              P= .001      P= .         P= .656      P= .125

VAR00003        .2928       -.1612       1.0000        .1180
              (    8)      (   10)      (   10)      (    9)
              P= .482      P= .656      P= .         P= .762

VAR00004        .6809       -.5501        .1180       1.0000
              (    7)      (    9)      (    9)      (    9)
              P= .092      P= .125      P= .762      P= .

(Coefficient / (Cases) / 2-tailed Significance)

" . " is printed if a coefficient cannot be computed
```

Table 15.3 **Exclude cases listwise output from Correlate**

-- Correlation Coefficients --

	VAR00001	VAR00002	VAR00003	VAR00004
VAR00001	1.0000	−.9564	.3536	.6809
	(7)	(7)	(7)	(7)
	P= .	P= .001	P= .437	P= .092
VAR00002	−.9564	1.0000	−.4830	−.7294
	(7)	(7)	(7)	(7)
	P= .001	P= .	P= .272	P= .063
VAR00003	.3536	−.4830	1.0000	.0875
	(7)	(7)	(7)	(7)
	P= .437	P= .272	P= .	P= .852
VAR00004	.6809	−.7294	.0875	1.0000
	(7)	(7)	(7)	(7)
	P= .092	P= .063	P= .852	P= .

(Coefficient / (Cases) / 2-tailed Significance)
'' . '' is printed if a coefficient cannot be computed

15.4 Interpreting the output in Tables 15.2 and 15.3

There is little in the output which has not been discussed in other chapters. The only thing to bear in mind is that the statistics are based on a reduced number of cases.

15.5 Reporting the output in Tables 15.2 and 15.3

Remember to report the actual sample sizes used in reporting each statistical analysis rather than the number of cases overall.

Chapter 16

Recoding values

Sometimes researchers need to alter how certain values of a variable are recorded by the computer; perhaps several different values need to be combined into one. The recoding values procedure allows you considerable flexibility to modify quickly and easily how any value has been coded numerically.

16.1 Introduction

Sometimes we need to recode values for a particular variable in our data. There can be many reasons for this, including:

1. To put together several categories of a nominal variable which otherwise have very few cases. This is commonly employed in statistics such as chi-square.
2. To place score variables into ranges of scores.

We may wish to categorise our sample into two or more groups according to some variable such as age or intelligence. We will illustrate the recoding of cases with the data in Table 16.1 which shows the music and mathematics scores of 10 children together with their code for gender and their age in years. The music and mathematics scores are the same as those previously presented in Tables 10.1 and 15.1. Suppose we wanted to compute the correlation between the music and mathematics scores for the younger and older children. To do this, we would first have to decide how many age groups we wanted. Since we have only 10 children we will settle for two groups. Next we decide what the cut-off point in age will be for the two groups. As we want two groups of similar size, we will select 10 as the cut-off point with children younger than 10 falling into one group and children aged 10 or more into the other group. We will now use SPSS to recode age in this way.

Table 16.1 **Scores on musical ability and mathematical ability for 10 children with their sex and age**

Music score	Mathematics score	Sex	Age
2	8	1	10
6	3	1	9
4	9	2	12
5	7	1	8
7	2	2	11
7	3	2	13
2	9	2	7
3	8	1	10
5	6	2	9
4	7	1	11

16.2 Recoding values

Quick summary

Enter data in Newdata

Transform

Recode

Into Different Variables . . .

Variable to be recoded and ▶

Name: box

Type name of recoded variable

Change

Old and New Values . . .

  Old value or range of old values

  Type old value

  Value: box

  Type new value

  Add

  Repeat as necessary

Continue

OK

	var00001	var00002	var00003	var00004	var
1	2.00	8.00	1.00	10.00	
2	6.00	3.00	1.00	9.00	
3	4.00	9.00	2.00	12.00	
4	5.00	7.00	1.00	8.00	
5	7.00	2.00	2.00	11.00	
6	7.00	3.00	2.00	13.00	
7	2.00	9.00	2.00	7.00	
8	3.00	8.00	1.00	10.00	
9	5.00	6.00	2.00	9.00	
10	4.00	7.00	1.00	11.00	
11					

Figure 16.1 **Music and mathematics scores with gender and age in Newdata**

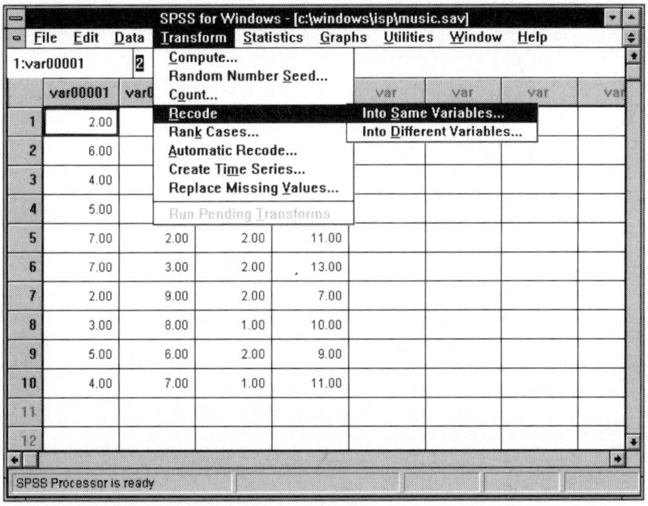

SPSS for Windows - [c:\windows\isp\music.sav]

File Edit Data Transform Statistics Graphs Utilities Window Help

1:var00001 2

Compute...
Random Number Seed...
Count...
Recode
Rank Cases...
Automatic Recode...
Create Time Series...
Replace Missing Values...
Run Pending Transforms

Into Same Variables...
Into Different Variables...

	var00001	var		
1	2.00			
2	6.00			
3	4.00			
4	5.00			
5	7.00	2.00	2.00	11.00
6	7.00	3.00	2.00	13.00
7	2.00	9.00	2.00	7.00
8	3.00	8.00	1.00	10.00
9	5.00	6.00	2.00	9.00
10	4.00	7.00	1.00	11.00
11				
12				

SPSS Processor is ready

Figure 16.2 **Transform and Recode drop-down menus**

- Enter the data in Table 16.1 in Newdata, putting music scores in the first column, mathematics scores in the second column, the code for gender in the third column and the age in years in the fourth column as shown in Figure 16.1. If you saved the data used in Chapter 10, then you can retrieve this data file.
- Select Transform from the menu bar in the Applications window which produces a drop-down menu (Figure 16.2).
- Select Recode which produces a second drop-down menu (Figure 16.2).
- Select Into Different Variables . . . which opens the Recode into Different Variables dialog box (Figure 16.3) and which creates a separate variable for you to store the recoded values. This is useful when you do not wish to lose the original values as in this case.
- Select var00004 (age) and the ▶ button which puts var00004 in the box called Numeric Variable –> Output Variable:.
- Select box under Name: and type in name for the recoded variable (e.g. 'agecat').
- Select Change.
- Select Old and New Values . . . which opens the Recode into Different Variables: Old and New Values dialog box (Figure 16.4).
- Select Range: in the Old Value section and type 9 in the box beside Lowest through.
- Select the box beside Value: in the New Value section and type 1.
- Select Add which puts Lowest thru 9 –> 1 in the Old –> New: box.
- Select Range: and type 10 in the box beside through highest.

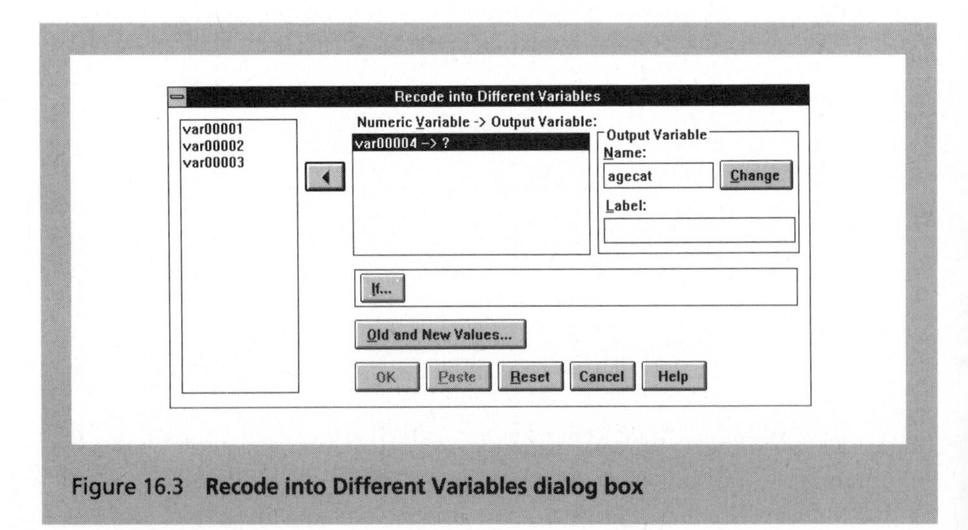

Figure 16.3 **Recode into Different Variables dialog box**

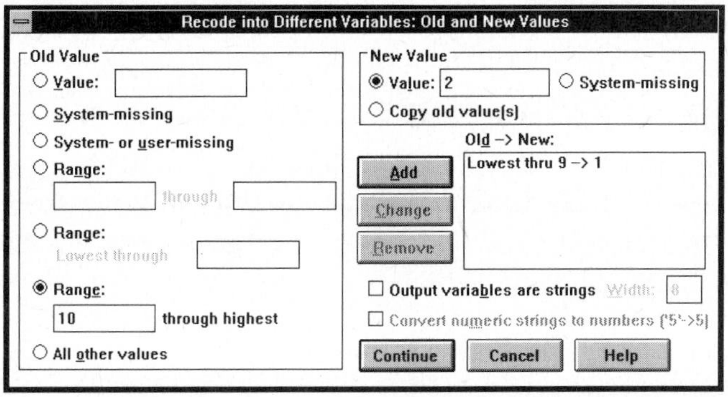

Figure 16.4 Recode into Different Variables: Old and New Values dialog box

	var00001	var00002	var00003	var00004	agecat	var
1	2.00	8.00	1.00	10.00	2.00	
2	6.00	3.00	1.00	9.00	1.00	
3	4.00	9.00	2.00	12.00	2.00	
4	5.00	7.00	1.00	8.00	1.00	
5	7.00	2.00	2.00	11.00	2.00	
6	7.00	3.00	2.00	13.00	2.00	
7	2.00	9.00	2.00	7.00	1.00	
8	3.00	8.00	1.00	10.00	2.00	
9	5.00	6.00	2.00	9.00	1.00	
10	4.00	7.00	1.00	11.00	2.00	
11						

Figure 16.5 Age recoded in Newdata

- Select the box beside Value: in the New Value section and type 2.
- Select Add which puts 10 thru Highest –> 2 in the Old –> New: box.
- Select Continue which closes the Recode into Different Variables: Old and New Values dialog box.

- Select OK which closes the Recode into Different Variables dialog box and which inserts the recoded values for age in the fifth column of Newdata called agecat as shown in Figure 16.5.

16.3 Reporting the output

With a complex set of data it is very easy to forget precisely what you have done to your data. Recoding can radically alter the output from a computer analysis. You need to carefully check the implications of any recodes before reporting them.

Chapter 17

Computing new variables

Computing new variables allows you to add, subtract, etc., scores on several variables to give you a new variable. For example, you might wish to add together several questions on a questionnaire to give an overall index of what the questionnaire is measuring.

17.1 Introduction

When analysing data we may want to form a new variable out of one or more old ones. For example, when measuring psychological variables several questions are often used to measure more or less the same thing. For instance, the following four statements might be used to assess satisfaction with life:

(a) I generally enjoy life

(b) Some days things just seem to get on top of me

(c) Life often seems pretty dull

(d) The future looks hopeful

Participants are asked to state how much they agree with each of these statements on the following four-point scale:

1: Strongly agree 2: Agree 3: Disagree 4: Strongly disagree

Table 17.1 **Life satisfaction scores of three respondents**

Respondent	(a) Enjoy life (recoded)	(b) On top of me	(c) Dull	(d) Hopeful (recoded)
1	Agree (3)	Agree (2)	St. disagree (4)	Agree (3)
2	Disagree (2)	Disagree (3)	Agree (2)	St. disagree (1)
3	St. agree (4)	Disagree (3)	Disagree (3)	Disagree (2)

We may use these four items to determine how satisfied people are with their life by adding up their responses to all four of them.

Notice a problem that frequently occurs when dealing with questionnaires: if you answer 'Strongly agree' to the first and fourth items you indicate that you enjoy life, while if you answer 'Strongly agree' to the second and third items you imply that you are dissatisfied with life. We want higher scores to denote greater life satisfaction. Consequently, we will reverse the scoring for the *first* and *fourth* items as follows:

1: Strongly disagree   2: Disagree   3: Agree   4: Strongly agree

We can use the Recode procedure described in Chapter 16 to recode the values for the first and fourth items.

The data in Table 17.1 show the answers to the four statements by three individuals in which the answers to the first and fourth items have been recoded numerically. We will use these data to illustrate the SPSS procedure for adding together the answers to the four statements to form an index of life satisfaction.

17.2 Computing a new variable

Quick summary

Transform

Compute . . .

Name of new variable

Basis of new variable

OK

	var00001	var00002	var00003	var00004	var
1	3.00	2.00	4.00	3.00	
2	2.00	3.00	2.00	1.00	
3	4.00	3.00	3.00	2.00	
4					

Figure 17.1 **Coded answers to four items in Newdata**

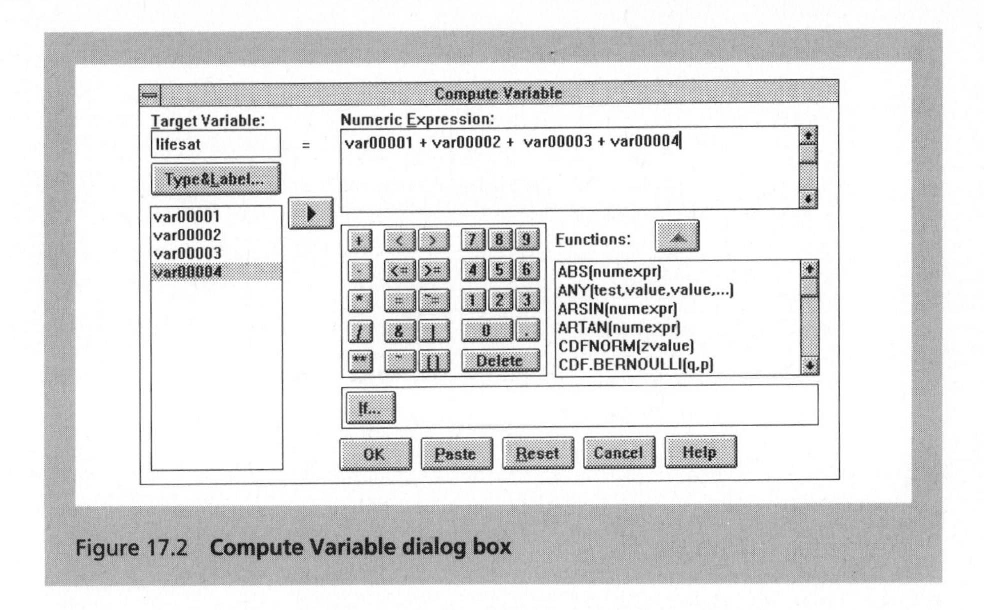

Figure 17.2 **Compute Variable dialog box**

3:lifesat 12

	var00001	var00002	var00003	var00004	lifesat	var
1	3.00	2.00	4.00	3.00	12.00	
2	2.00	3.00	2.00	1.00	8.00	
3	4.00	3.00	3.00	2.00	12.00	
4						

Figure 17.3 **Computed new variable in Newdata**

- Enter the data in Table 17.1 in Newdata as shown in Figure 17.1.
- Select Transform from the menu bar in the Applications window which produces a drop-down menu (Figure 16.2).
- Select Compute . . . which opens the Compute Variable dialog box (Figure 17.2).
- Type the name of the new variable (e.g. 'lifesat') in the Target Variable: box.
- Select var00001 and the ▶ button which puts var00001 in the Numeric Expression: box.
- Select + which puts + in the Numeric Expression: box.
- Repeat this procedure for the var00002, var00003 and var00004.

- Select OK which closes the Compute Variable dialog box and returns to the Newdata window displaying the new variable and its values as shown in Figure 17.3.

Do not forget to save your newly computed variables along with the rest of the spreadsheet Newdata on exiting SPSS if you are likely to want to use them again.

Chapter 18

Ranking tests

Nonparametric statistics

Sometimes you may wish to know whether the 'means' of two different sets of scores are significantly different from each other but feel that the requirement that the scores on each variable are roughly normally distributed (bell-shaped) is not fulfilled. Nonparametric tests can be used in these circumstances.

We will illustrate the computation of two nonparametric tests for related scores with the data in Table 18.1 which was also used in Chapter 12 and which shows the number of eye-contacts made by the same babies with their mothers at six and nine months. Notice that the Sign test (Section 18.1) and Wilcoxon Matched Pairs test (Section 18.4) produce different significance levels. The Sign test seems rather less powerful at detecting differences than the Wilcoxon Matched Pairs test.

Table 18.1 **Number of one-minute segments with eye-contact by babies at six and nine months**

	Six months	**Nine months**
Baby Clara	3	7
Baby Martin	5	6
Baby Sally	5	3
Baby Angie	4	8
Baby Trevor	3	5
Baby Sam	7	9
Baby Bobby	8	7
Baby Sid	7	9

18.1 Related scores: Sign test

Quick summary

Retrieve/enter data

Statistics

Nonparametric Tests

2 Related Samples . . .

var00001

var00002

▶

Wilcoxon

Sign

OK

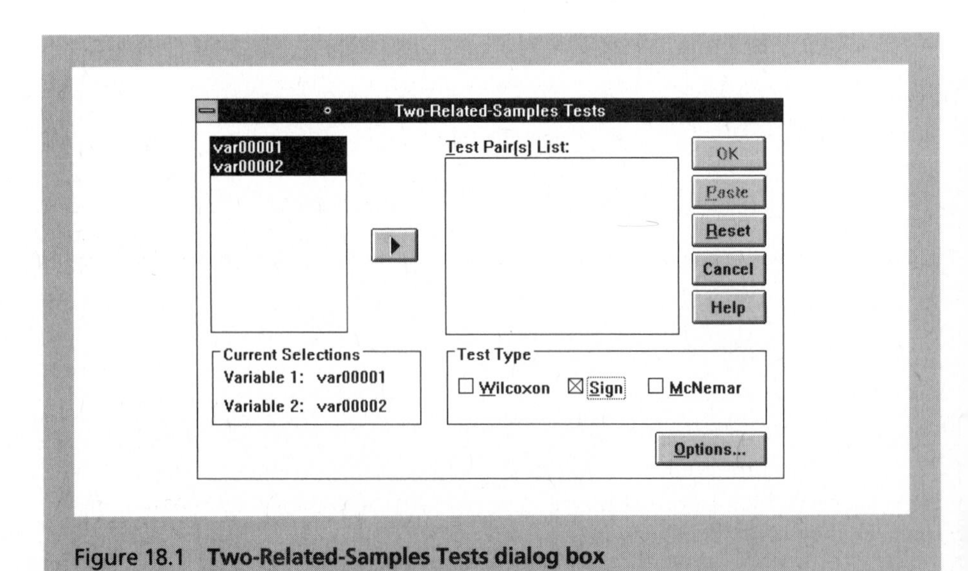

Figure 18.1 Two-Related-Samples Tests dialog box

- If you saved the data set used in Chapter 12, retrieve it. Otherwise create it again.
- Select Statistics on the menu bar of the Applications window which produces the drop-down menu shown in Figure 2.6.
- Select Nonparametric Tests from this drop-down menu which opens a second drop-down menu.
- Select 2 Related Samples . . . which opens the Two-Related-Samples Tests dialog box shown in Figure 18.1.
- Select 'var00001' which puts it beside Variable 1: in the Current Selections box.
- Select 'var00002' which puts it beside Variable 2: in the Current Selections box.
- Select the ▶ button which puts 'var00001-var00002' in the Test Pair[s] List: box.
- (If you wish) select Wilcoxon to de-select it.
- Select Sign.
- Select OK which closes the Two-Related-Samples Tests dialog box and the Newdata window and which displays the output shown in Table 18.2 in the Output window.

Table 18.2 **Sign test output**

```
-----Sign Test

     VAR00001
with VAR00002

  Cases

      2 -Diffs (VAR00002 LT VAR00001)
      6 +Diffs (VAR00002 GT VAR00001)   (Binomial)
      0  Ties                            2-Tailed P =   .2891
      -
      8  Total
```

18.2 Interpreting the output in Table 18.2

- This is straightforward. There are two negative signed differences and six positive signed differences. There were no ties, so the number of cases is eight.
- The two-tailed probability is 0.2891 or 29%, which is clearly not significant at the 5% level. (Binomial) refers to the statistical technique by which probabilities can be found for samples consisting of just two different possible values, as is the case with the sign test (given that we ignore ties).

18.3 Reporting the output in Table 18.2

We could report these results as follows: 'There was no significant change in the amount of eye-contact between 6 and 9 months (Sign test, $p > 0.05$).'.

18.4 Related scores: Wilcoxon test

As Wilcoxon is the default option on the Two-Related-Samples Tests dialog box, simply select OK which closes the Two-Related-Samples Tests dialog box and the Newdata window and which displays the output shown in Table 18.3 in the Output window.

Table 18.3 **Wilcoxon test output**

```
-----Wilcoxon Matched-Pairs Signed-Ranks Test

    VAR00001
with VAR00002

  Mean Rank    Cases

      3.00         2  -Ranks (VAR00002 LT VAR00001)
      5.00         6  +Ranks (VAR00002 GT VAR00001)
                   0   Ties  (VAR00002 EQ VAR00001)
                   -
                   8   Total

       Z = -1.6803             2-Tailed P = .0929
```

18.5 Interpreting the output in Table 18.3

- The output tells us that there are two cases which were negatively signed after ranking and six cases which were positively signed after ranking. It seems clear that VAR00002 tends to have larger values than VAR00001.
- Instead of using tables of critical values, the computer uses a formula which relates to the z-distribution. The z-value is –1.6803 which has a two-tailed probability of 0.0929. This means that the difference between the two variables is not statistically significant at the 5% level.

18.6 Reporting the output in Table 18.3

We could report these results as follows: 'There was no significant difference in the amount of eye-contact by babies between 6 and 9 months (Wilcoxon, $z = -1.68$, $p > 0.05$).'.

Table 18.4 **Emotionality scores in two-parent and lone-parent families**

Two-parent family	Lone-parent family
12	6
18	9
14	4
10	13
19	14
8	9
15	8
11	12
10	11
13	9
15	
16	

18.7 Unrelated scores: Mann–Whitney U-test

We will illustrate the computation of one nonparametric test for unrelated scores with the data in Table 18.4 which shows the emotionality scores of 12 children from two-parent families and 10 children from single-parent families.

Quick summary

Retrieve/enter data

Statistics

Nonparametric Tests

2 Independent Samples . . .

Dependent or test variable (e.g. var00002)

▶

Grouping variable (e.g. var00001)

▶

Define Groups . . .

1

2

Continue

OK

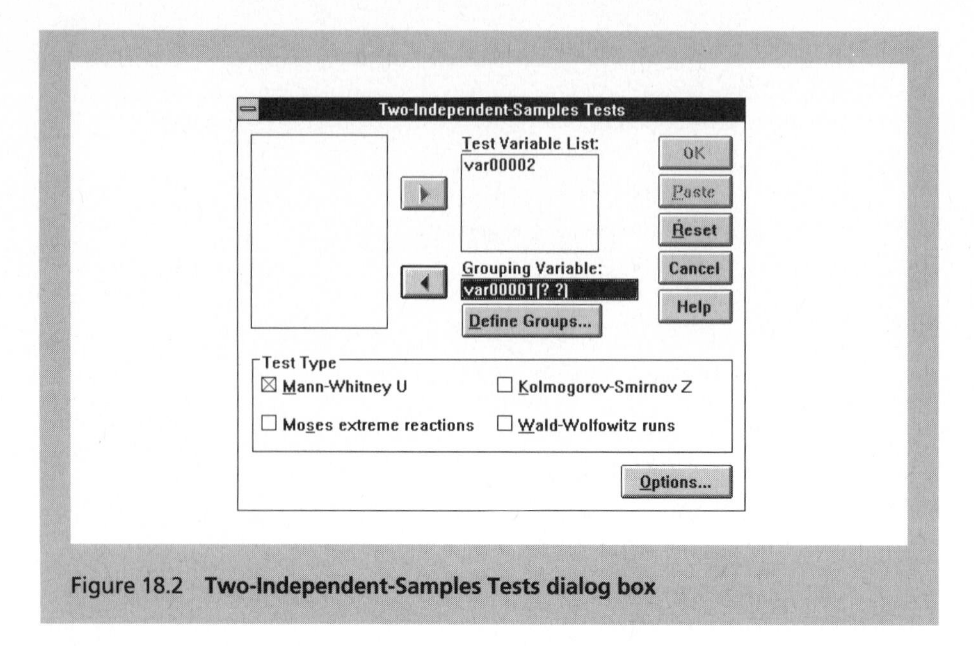

Figure 18.2 **Two-Independent-Samples Tests dialog box**

Two Independent Samples: Define Groups

Group 1: 1

Group 2: 2

Continue

Cancel

Help

Figure 18.3 **Two Independent Samples: Define Groups dialog box**

- If you saved the data set used in Chapter 13, retrieve it. Otherwise create it again.
- Select Statistics on the menu bar of the Applications window which produces the drop-down menu shown in Figure 2.6.
- Select Nonparametric Tests from this drop-down menu which opens a second drop-down menu.
- Select 2 Independent Samples . . . which opens the Two-Independent-Samples Tests dialog box shown in Figure 18.2.
- Select 'var00002' and the ▶ button beside Test Variable List: which puts 'var00002' in this box.

- Select 'var00001' and the ▶ button beside Grouping Variable: which puts 'var00001' in this box.
- Select Define Groups . . . which opens the Two Independent Samples: Define Groups dialog box shown in Figure 18.3.
- Type 1 in the box beside Group 1:.
- Select box beside Group 2: and type 2.
- Select Continue which closes the Two Independent Samples: Define Groups dialog box.
- Select OK which closes the Two-Independent-Samples Tests dialog box and the Newdata window and which displays the output shown in Table 18.5.

Table 18.5 **Mann–Whitney U-test output**

```
-----Mann–Whitney U – Wilcoxon Rank Sum W Test

  VAR00002
by VAR00001

  Mean Rank  Cases
       7.85     10  VAR00001 = 1.00
      14.54     12  VAR00001 = 2.00
                 –
                22  Total

                              Exact          Corrected for ties
        U          W        2-Tailed P          Z        2-Tailed P
      23.5       78.5         .0138         -2.4142        .0158
```

18.8 Interpreting the output in Table 18.5

- The table indicates that the average rank given to VAR00001 for the first group (i.e. value = 1.00) is 7.85, and the average rank given to the second group (i.e. value = 2.00) is 14.54. This means that the scores for Group 2 tend to be larger than those for Group 1.
- The basic Mann–Whitney statistic, the U-value, is 23.5, which is statistically significant at the 0.0138 level.
- In addition, the computer has printed out a Z value of –2.4142 which is significant at the 0.0158 level. This is the value of the Mann–Whitney test when a correction for tied ranks has been applied. As can be seen, this has altered the significance level only marginally to 0.0158 from 0.0138.

18.9 Reporting the output in Table 18.5

We could report the results of this analysis as follows: 'The Mann–Whitney U-test found that the emotionality scores of children from two-parent families were significantly higher than those of children in lone-parent families ($U = 23.5$, $p = 0.014$).'.

Chapter 19

The variance ratio test

The *F*-ratio to compare two variances

The variance ratio test (F-test) indicates whether two sets of scores differ in the variability of the scores around the mean (i.e. whether the variances are significantly different).

19.1 Introduction

To compute the variance or *F*-ratio we divide the larger variance estimate by the smaller variance estimate. The variance estimate is produced by the Describe procedure which we first introduced in Chapter 5. We will illustrate the computation of the variance ratio with the data in Table 19.1 (*ISP* Table 19.2) which reports the emotional stability scores of patients who have had an electric current passed through either the left or the right hemisphere of the brain. In this chapter, the method involves a little hand-calculation. However, it provides extra experience with SPSS.

Table 19.1 **Emotional stability scores from a study of ECT to different hemispheres of the brain**

Left hemisphere	Right hemisphere
20	36
14	28
18	4
22	18
13	2
15	22
9	1
Mean = 15.9	Mean = 15.9

An alternative way of achieving the same end is to follow the *t*-test procedures in Chapter 13. You may recall that the Levene *F*-ratio test is part of the output for that *t*-test. Although Levene's test is slightly different, it is a useful alternative to the conventional *F*-ratio test.

19.2 Variance estimate

Quick summary

Enter data

Statistics

Summarize

Descriptives . . .

Select variables ▶

Options . . .

Variance

Continue

OK

- Enter the data in Table 19.1 in Newdata, putting the emotional stability scores in the first column for the left hemisphere condition and in the second column for the right hemisphere condition.
- Select Statistics which produces a drop-down menu (Figure 2.6).
- Select Summarize which displays a second drop-down menu (also Figure 2.6).
- Select Descriptives . . . which opens the Descriptives dialog box (Figure 5.1).
- Select 'var00001' and then the ▶ button which puts 'var00001' in the Variable[s]: text box.
- Select 'var00002' and then the ▶ button which puts 'var00002' in the Variable[s]: text box.
- Select Options . . . which opens the Descriptives: Options sub-dialog box (Figure 5.2).
- Select Variance.
- Select Continue which closes the Descriptives: Options sub-dialog box.
- Select OK which closes the Descriptives dialog box and Newdata window and which displays the output shown in Table 19.2 in the Output window.

Table 19.2 **Estimated variance and the default descriptive statistics produced by the Descriptives procedure**

Number of valid observations (listwise) = 7.00

Variable Label	Mean	Std Dev	Variance	Minimum	Maximum	Valid N
VAR00001	15.86	4.45	19.81	9.00	22.00	7
VAR00002	15.86	13.84	191.48	1.00	36.00	7

19.3 Interpreting the output in Table 19.2

We compute the variance or *F*-ratio from variance estimates as follows:

- Divide the larger variance estimate of Table 19.2 by the smaller variance estimate. The larger variance estimate is 191.48 (for 'var00002'), which divided by the smaller one of 19.81 (for 'var00001') gives a variance or *F*-ratio of 9.6658. This ratio is 9.66 when rounded down to two decimal places.
- We need to look up the statistical significance of this ratio in a table of critical values of *F*-ratios where the degrees of freedom for the numerator (191.48) and the denominator (19.81) of the ratio are both 6 (see *ISP* Significance Table 19.1).
- The 0.05 (5%) critical value of the *F*-ratio with 6 degrees of freedom in the numerator and denominator is 4.28.
- The *F*-ratio we obtained is 9.66 which is larger than the 0.05 critical value of 4.28.
- Consequently, we would conclude that the variance of emotionality scores of patients in the right hemisphere condition was significantly larger than that of patients in the left hemisphere condition.

Chapter 20

Analysis of variance (ANOVA)

Introduction to the one-way unrelated or uncorrelated ANOVA

The unrelated/uncorrelated analysis of variance tells you whether several (two or more) sets of scores have very different means. It assumes that the sets of scores come from different individuals.

We will illustrate the computation of a one-way unrelated analysis of variance with the data in Table 20.1 (*ISP* Table 20.2) which shows the scores of different participants in three conditions. It is a study of the effect of different hormone and placebo treatments on depression. So drug is the independent variable and depression the dependent variable.

Table 20.1 **Data for a study of the effects of hormones**

Group 1 Hormone 1	Group 2 Hormone 2	Group 3 Placebo control
9	4	3
12	2	6
8	5	3

20.1 One-way unrelated ANOVA

Quick summary

Enter data

Statistics

Compare Means

One-Way ANOVA . . .

Dependent variable (e.g. var00002)

▶

Independent or factor variable (e.g. var00001)

▶

Define Range . . .

Minimum group code (e.g. 1)

Maximum:

Maximum group code (e.g. 3)

Continue

Options . . .

Descriptive

Continue

OK

---

	var00001	var00002	var
1	1.00	9.00	
2	1.00	12.00	
3	1.00	8.00	
4	2.00	4.00	
5	2.00	2.00	
6	2.00	5.00	
7	3.00	3.00	
8	3.00	6.00	
9	3.00	3.00	
10			

Figure 20.1 Data for a one-way unrelated ANOVA in Newdata

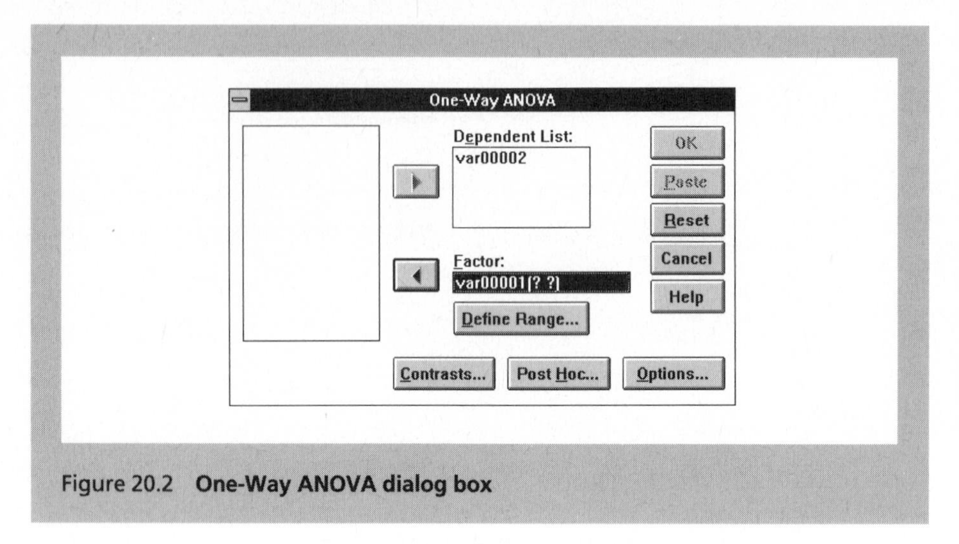

Figure 20.2 **One-Way ANOVA dialog box**

Figure 20.3 **One-Way ANOVA: Define Range dialog box**

- Enter the data in Table 20.1 in Newdata, putting the code for the three conditions (e.g. 1 for Group 1, 2 for Group 2, and 3 for Group 3) in the first column and the score for each of the participants in those three conditions in the second column as shown in Figure 20.1.
- Select Statistics on the menu bar of the Applications window which produces a drop-down menu (Figure 2.6).
- Select Compare Means from this drop-down menu which opens a second drop-down menu.
- Select One-Way ANOVA . . . which opens the One-Way ANOVA dialog box (Figure 20.2).
- Select 'var00002' and the ▶ button beside Dependent List: which puts 'var00002' in this box.
- Select 'var00001' and the ▶ button beside Factor: which puts 'var00001' in this box.

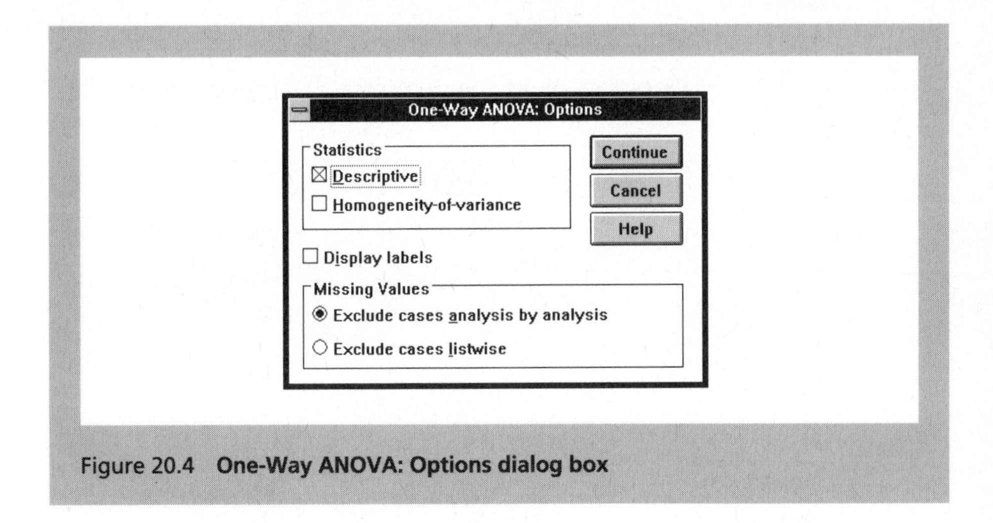

Figure 20.4 **One-Way ANOVA: Options dialog box**

- Select Define Range . . . which opens the One-Way ANOVA: Define Range dialog box (Figure 20.3).
- Type 1 in the box beside Minimum:.
- Select box beside Maximum: and type 3.
- Select Continue which closes the One-Way ANOVA: Define Range dialog box.
- Select Options . . . which opens the One-Way ANOVA: Options dialog box shown in Figure 20.4.
- Select Descriptive.
- Select Continue which closes the One-Way ANOVA: Options dialog box.
- Select OK which closes the One-Way ANOVA dialog box and the Newdata window and which displays the output shown in Table 20.2 (p.136).

20.2 Interpreting the output in Table 20.2

- The analysis of variance table is presented in the upper part of the output while the descriptive statistics are displayed in the lower part of the output.
- The F-ratio is the Between Groups Mean Square divided by the Within Groups Mean Square which gives an F-ratio of 10.5862 (34.1111/3.2222 = 10.5862).
- The probability of this F-ratio is 0.0108. In other words, it is less than the 0.05 critical value and so is statistically significant.
- This indicates that there is a significant difference between the three groups. *However, it does not necessarily imply that all the means are significantly different from each other. In this case, one suspects that the means 3.6667 and 4.0000 are not significantly different.*

Table 20.2 **One-way ANOVA output**

```
-----O N E   W A Y-----

  Variable VAR00002
By Variable VAR00001

                    Analysis of Variance

                              Sum of        Mean           F         F
Source               D.F.    Squares      Squares       Ratio     Prob.

Between Groups          2    68.2222      34.1111     10.5862     .0108
Within Groups           6    19.3333       3.2222
Total                   8    87.5556

                     Standard Standard
Group  Count    Mean Deviation    Error  95 Pct Conf  Int for Mean

Grp 1      3  9.6667    2.0817   1.2019   4.4955  TO      14.8379
Grp 2      3  3.6667    1.5275    .8819   -.1280  TO       7.4613
Grp 3      3  4.0000    1.7321   1.0000   -.3027  TO       8.3027

Total      9  5.7778    3.3082   1.1027   3.2348  TO       8.3207

GROUP        MINIMUM        MAXIMUM

Grp 1         8.0000        12.0000
Grp 2         2.0000         5.0000
Grp 3         3.0000         6.0000

TOTAL         2.0000        12.0000
```

- Which of the means differ from the others can be further determined by the use of multiple comparison tests such as the unrelated *t*-test. To do this, minimise your output to take you back to Newdata. Then follow the procedure for the unrelated *t*-test described in Chapter 13. You do not have to re-enter your data. However, do an unrelated *t*-test defining the groups as 1 and 2, then redefine the groups as 1 and 3 (you will find the screen locked until you select reset at this stage), and finally redefine the groups as 2 and 3 (remembering to press reset). For our example, group 1 is significantly different from groups 2 and 3 which do not differ significantly from each other. (See *ISP* Chapter 13 for more details.)

20.3 Reporting the output in Table 20.2

We could report the results of the output as follows: 'The effect of the drug treatment was significant overall ($F_{2,6} = 10.58$, $p<0.05$). However, the mean for hormone treatment 1 seems rather different from those for the other two groups.'.

Chapter 21

Analysis of variance for correlated scores or repeated measures

The correlated/related analysis of variance tells you whether several (two or more) sets of scores have very different means. However, it assumes that a single sample of individuals have contributed scores to each of the different sets of scores and that the correlation coefficients between sets of scores are large.

We will illustrate the computation of a one-way correlated analysis of variance with the data in Table 21.1 which shows the scores of the same participants in three different conditions (*ISP* Table 21.10).

Table 21.1 **Pain relief scores from a drugs experiment**

	Aspirin	'Product X'	Placebo
Bob Robertson	7	8	6
Mavis Fletcher	5	10	3
Bob Polansky	6	6	4
Ann Harrison	9	9	2
Bert Entwistle	3	7	5

21.1 One-way correlated ANOVA

Quick summary

Enter data

Statistics

ANOVA Models

Repeated Measures . . .

Number of levels of factor (e.g. 3)

Add

Define

First variable/level (e.g. 'var00001')

▶ *button beside Within-Subjects Variables:*

Repeat as necessary

OK

Obtain the means and standard deviations of each variable (Statistics, Summarize, Descriptives etc.: Chapter 5)

- Enter the data in Table 21.1 in Newdata, putting the score for the first condition (aspirin) in the first column, the score for the second condition (product X) in the second column and the score for the third condition (placebo) in the third column as shown in Figure 21.1.
- Select Statistics on the menu bar of the Applications window which produces a drop-down menu (Figure 2.6).
- Select ANOVA Models which produces a second drop-down menu.
- Select Repeated Measures . . . which opens the Repeated Measures Define Factor(s) dialog box (Figure 21.2).
- Type 3 in the box beside Number of Levels: and select Add which puts the expression 'factor1(3)' in the bottom box.
- Select Define which opens the Repeated Measures ANOVA dialog box (Figure 21.3).

	var00001	var00002	var00003	var
1	7.00	8.00	6.00	
2	5.00	10.00	3.00	
3	6.00	6.00	4.00	
4	9.00	9.00	2.00	
5	3.00	7.00	5.00	
6				

Figure 21.1 Data for one-way correlated ANOVA in Newdata

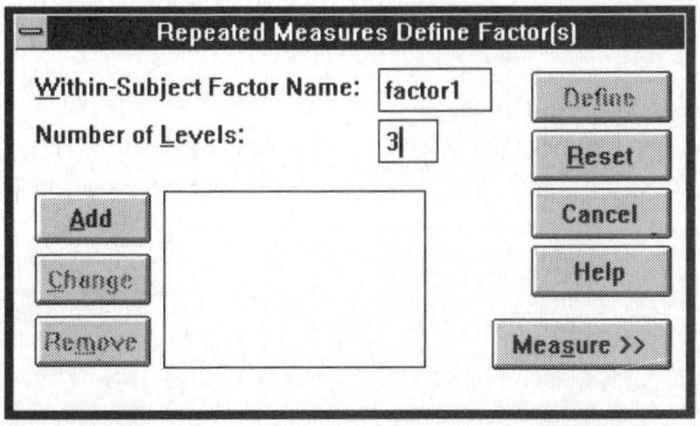

Figure 21.2 **Repeated Measures Define Factor(s) dialog box**

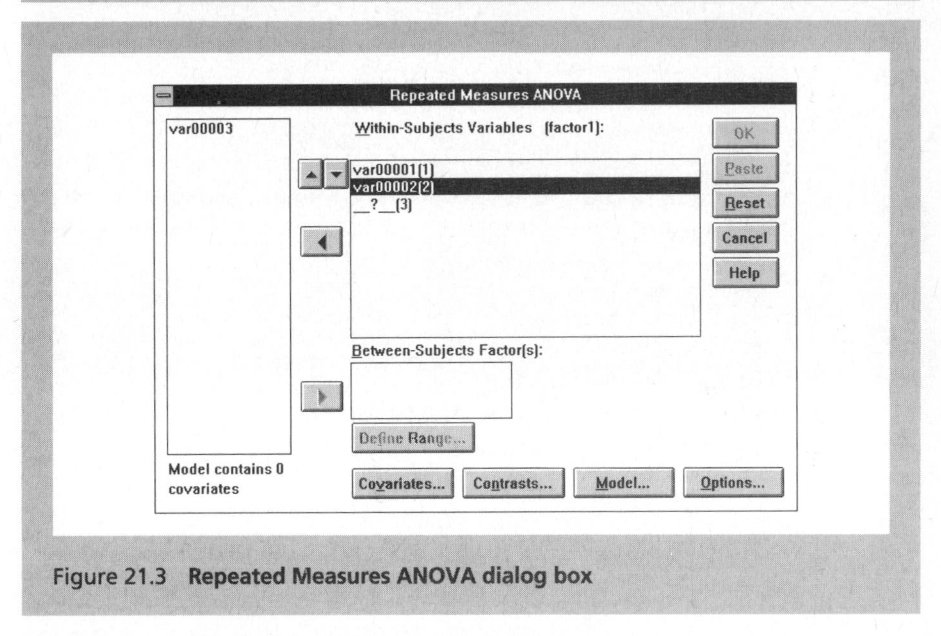

Figure 21.3 **Repeated Measures ANOVA dialog box**

- Select 'var00001' and the ▶ button beside Within-Subjects Variables [factor1]: which puts 'var00001' in this box.
- Repeat this procedure with 'var00002' and 'var00003'.
- Select OK which closes the Repeated Measures ANOVA dialog box and the Newdata window and which presents the output shown in Table 21.2 in the Output window.
- *It is also vital to know the mean of each condition when you interpret the ANOVA output*. To display the means and standard deviations for the three conditions use the Descriptives procedure first described in Chapter 5: Statistics, Summarize, Descriptives, etc. (p. 38).

Table 21.2 **One-way correlated ANOVA output**

```
* * * * * * Analysis of Variance * * * * * *

     5 cases accepted.
     0 cases rejected because of out-of-range factor values.
     0 cases rejected because of missing data.
     1 non-empty cell.
     1 design will be processed.
- - - - - - - - - - - - - - - - - - - - - - - - - - - - - - - - - - -
* * * * * * Analysis of Variance - design 1 * * * * * *

Tests of Between-Subjects Effects.

Tests of Significance for T1 using UNIQUE sums of squares
Source of Variation          SS      DF        MS          F    Sig of F

WITHIN+RESIDUAL            8.67       4      2.17
CONSTANT                 540.00       1    540.00     249.23        .000
- - - - - - - - - - - - - - - - - - - - - - - - - - - - - - - - - - -
* * * * * * Analysis of Variance - design 1 * * * * * *

Tests involving 'FACTOR1' Within-Subject Effect.

Mauchly sphericity test, W =    .86238
Chi-square approx. =            .44417 with 2 D. F.
Significance =                    .801

Greenhouse-Geisser Epsilon =    .87903
Huynh-Feldt Epsilon =          1.00000
Lower-bound Epsilon =           .50000

AVERAGED Tests of Significance that follow multivariate tests are equi-
valent to univariate or split-plot or mixed-model approach to repeated
measures. Epsilons may be used to adjust d.f. for the AVERAGED results.
- - - - - - - - - - - - - - - - - - - - - - - - - - - - - - - - - - -
* * * * * * Analysis of Variance - design 1 * * * * * *

EFFECT .. FACTOR1
Multivariate Tests of Significance (S = 1, M = 0, N = 1/2)

Test Name        Value    Exact F    Hypoth. DF    Error DF    Sig. of F

Pillais         .75530    4.62992          2.00        3.00         .121
Hotellings     3.08661    4.62992          2.00        3.00         .121
Wilks           .24470    4.62992          2.00        3.00         .121
Roys            .75530
Note.. F statistics are exact.
- - - - - - - - - - - - - - - - - - - - - - - - - - - - - - - - - - -
* * * * * * Analysis of Variance - design 1 * * * * * *

Tests involving 'FACTOR1' Within-Subject Effect.

AVERAGED Tests of Significance for VAR using UNIQUE sums of squares
Source of Variation          SS      DF        MS          F    Sig of F

WITHIN+RESIDUAL           31.33       8      3.92
FACTOR1                   40.00       2     20.00       5.11        .037
```

21.2 Interpreting the output in Table 21.2

- As can be seen SPSS produces a great deal of output for the Repeated Measures procedure. The output which is of most interest to us is the last table.
- The source of variation called 'WITHIN+RESIDUAL' is more commonly known as the Residual or Residual Error variance.
- The *F*-ratio is the Mean Square (MS) for 'FACTOR1' divided by Residual Mean Square. It is 5.10 (20.00/3.92 = 5.10).
- The exact significance level of this *F*-ratio is 0.037. Since this value is smaller than 0.05, we would conclude that there is a significant difference in the mean scores of the three conditions.
- In order to interpret the meaning of the ANOVA as it applies to your data, you need to consider the means of each of the three groups of scores. To repeat, these are not printed out by the ANOVA procedure. The Descriptives procedure first described in Chapter 5 will give you the means (Statistics, Summarize, Descriptives, etc.). They are 6.00, 8.00, and 4.00.
- You also need to remember that if you have three or more groups, you need to check where the significant differences lie between the pairs of groups. The related *t*-test procedure in Chapter 12 explains this. For the present example, only the difference between the means for product X and the placebo was significant. (See also Chapter 23.)

21.3 Reporting the output in Table 21.2

- We could describe the results of this analysis in the following way: 'A one-way correlated analysis of variance showed a significant treatment effect for the three conditions ($F_{2,8} = 5.10, p<0.05$). The Aspirin mean was 6.00, the Product X mean was 8.00, and the Placebo mean was 4.00. *t*-Tests between all three pairs suggested that there were no significant differences except for Product X versus the Placebo (t = 3.26, df = 4, p<0.05).'.
- This could be supplemented by an Analysis of Variance summary table such as Table 21.3. Drugs is FACTOR1 in the output, and Residual Error is WITHIN+RESIDUAL from the final table in the output (Table 21.2).

Table 21.3 **Analysis of Variance summary table**

Source of variation	Sum of squares	Degrees of freedom	Mean square	*F*-ratio
Drugs	40.00	2	20.00	5.10*
Residual error	31.33	8	3.92	–

* Significant at 5% level.

Chapter 22

Two-way analysis of variance for unrelated/uncorrelated scores

Two-way analysis of variance allows you to compare the means on the dependent variable when you have TWO independent variables.

We will illustrate the computation of a two-way unrelated analysis of variance with the data in Table 22.1 which shows the scores of different participants in six conditions reflecting the two factors of sleep deprivation and alcohol (*ISP* Table 22.11).

Table 22.1 **Data for sleep deprivation experiment: number of mistakes on video test**

	Sleep deprivation		
	4 hours	**12 hours**	**24 hours**
Alcohol	16	18	22
	12	16	24
	17	25	32
No alcohol	11	13	12
	9	8	14
	12	11	12

22.1 Two-way unrelated ANOVA

Quick summary

Enter data

Statistics

ANOVA Models

Simple Factorial . . .

Dependent variable and ▶ button beside Dependent:

Independent variable and ▶ button beside Factor[s]:

Define Range . . .

Minimum group code (e.g. 1)

Maximum:

Maximum group code (e.g. 3)

Continue

Repeat as necessary

Options . . .

Hierarchical or Experimental

Means and counts

Continue

OK

---

	var00001	var00002	var00003	var
1	1.00	1.00	16.00	
2	1.00	1.00	12.00	
3	1.00	1.00	17.00	
4	2.00	1.00	18.00	
5	2.00	1.00	16.00	
6	2.00	1.00	25.00	
7	3.00	1.00	22.00	
8	3.00	1.00	24.00	
9	3.00	1.00	32.00	
10	1.00	2.00	11.00	
11	1.00	2.00	9.00	

Figure 22.1 **Data for a two-way unrelated ANOVA in Newdata**

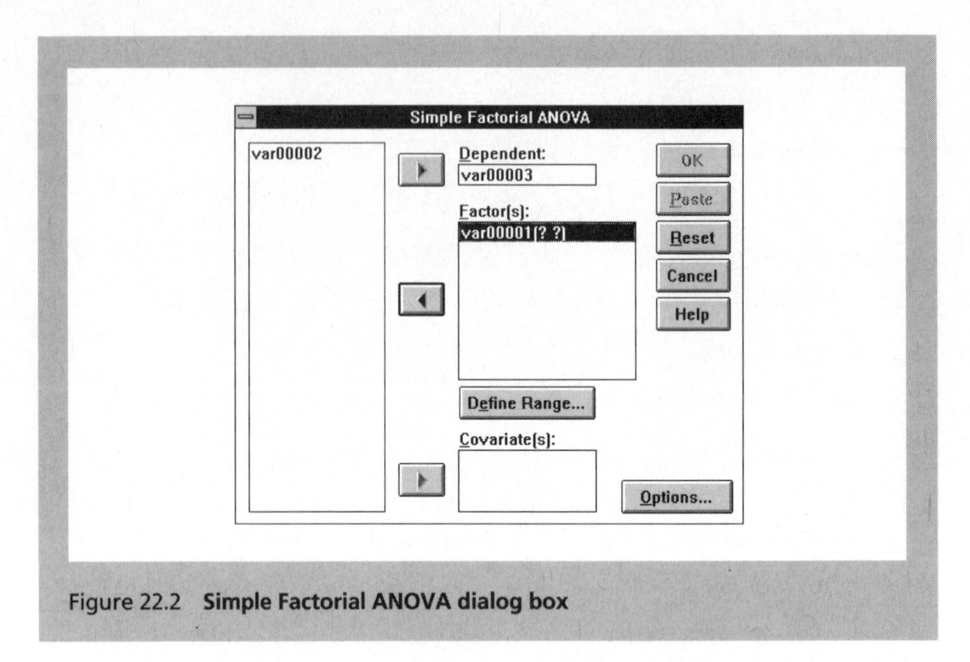

Figure 22.2 **Simple Factorial ANOVA dialog box**

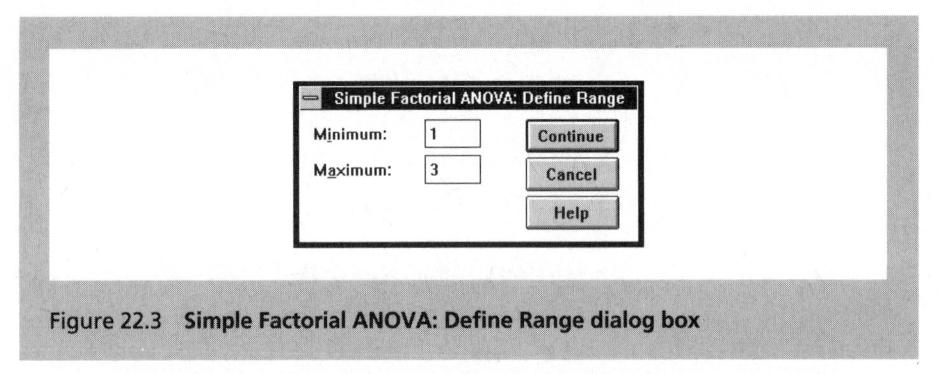

Figure 22.3 **Simple Factorial ANOVA: Define Range dialog box**

- Enter the data in Table 22.1 in Newdata, putting the code for the three sleep deprivation conditions (e.g. 1 for 4 hours, 2 for 12 hours, and 3 for 24 hours) in the first column, the code for the two alcohol conditions (e.g. 1 for alcohol and 2 for no alcohol) in the second column and the score for each participant in those six conditions in the third column as shown in Figure 22.1.
- Select Statistics on the menu bar of the Applications window which produces a drop-down menu (Figure 2.6).
- Select ANOVA Models from this drop-down menu which opens a second drop-down menu.
- Select Simple Factorial . . . which opens the Simple Factorial ANOVA dialog box (Figure 22.2).

- Select 'var00003' and the ▶ button beside Dependent: which puts 'var00003' in this box.
- Select 'var00001' and the ▶ button beside Factor[s]: which puts 'var00001' in this box.
- Select Define Range . . . which opens the Simple Factorial ANOVA: Define Range dialog box (Figure 22.3).
- Type 1 in the box beside Minimum:.
- Select box beside Maximum: and type 3.
- Select Continue which closes the Simple Factorial ANOVA: Define Range dialog box.
- Select 'var00002' and the ▶ button beside Factor[s]: which puts 'var00002' in this box.
- Select Define Range . . . again which opens the Simple Factorial ANOVA: Define Range dialog box.
- Type 1 in the box beside Minimum:.
- Select box beside Maximum: and type 2.
- Select Continue which closes the Simple Factorial ANOVA: Define Range dialog box.
- Select Options . . . which opens the Simple Factorial ANOVA: Options dialog box (Figure 22.4).
- Select either Hierarchical or Experimental in the Method section.
- Select Means and counts in the Statistics section.
- Select Continue which closes the Simple Factorial ANOVA: Options dialog box.
- Select OK which closes the Simple Factorial ANOVA dialog box and the Newdata window and which displays the output shown in Table 22.2.

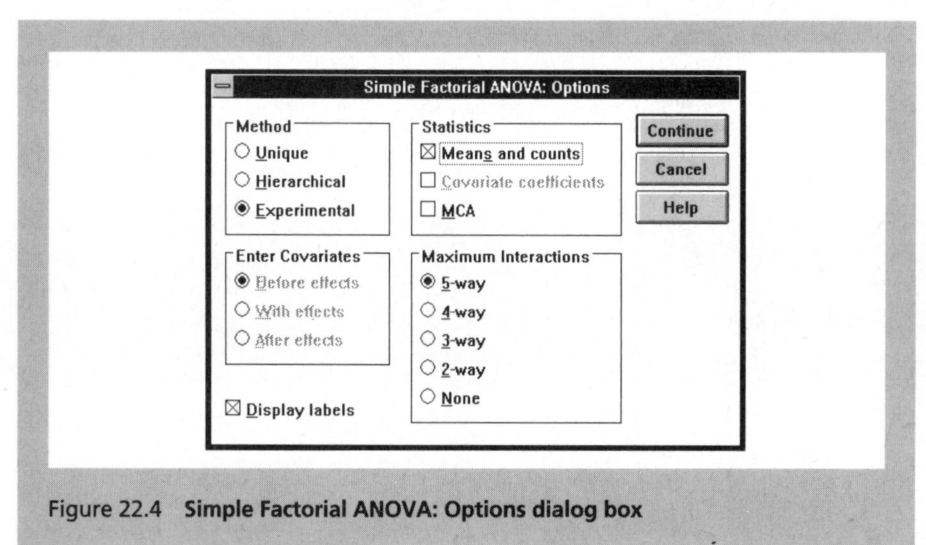

Figure 22.4 **Simple Factorial ANOVA: Options dialog box**

Table 22.2 **Two-way unrelated ANOVA output**

```
                    * * * C E L L   M E A N S * * *
      VAR00003
   by VAR00001
      VAR00002
Total Population
    15.78
  (   18)
VAR00001
        1           2           3
    12.83       15.17       19.33
  (    6)     (    6)     (    6)
VAR00002
        1           2
    20.22       11.33
  (    9)     (    9)

            VAR00002
VAR00001        1           2
        1   15.00       10.67
          (    3)     (    3)
        2   19.67       10.67
          (    3)     (    3)
        3   26.00       12.67
          (    3)     (    3)

          * * * A N A L Y S I S   O F   V A R I A N C E * * *
      VAR00003
   by VAR00001
      VAR00002
      EXPERIMENTAL sums of squares
      Covariates entered FIRST
                             Sum of             Mean               Sig
Source of Variation         Squares    DF     Square         F     of F
Main Effects                485.667     3    161.889    14.426     .000
  VAR00001                  130.111     2     65.056     5.797     .017
  VAR00002                  355.556     1    355.556    31.683     .000
2-Way Interactions           60.778     2     30.389     2.708     .107
  VAR00001 VAR00002          60.778     2     30.389     2.708     .107
Explained                   546.444     5    109.289     9.739     .001
Residual                    134.667    12     11.222
Total                       681.111    17     40.065
18 cases were processed.
0 cases (.0 pct) were missing.
```

22.2 Interpreting the output in Table 22.2

- The descriptive statistics are displayed in the upper part of the output while the Analysis of Variance table is presented in the lower part of the output.
- The mean for the total sample is shown first, followed by the means for the three conditions of the first variable (var00001), the means for the two conditions of the second variable (var00002) and the means for the six conditions comprising the two variables.
- Note that the standard deviations are not presented. To obtain these we need to use the Descriptives procedure first described in Chapter 5.
- In the Analysis of Variance table the *F*-ratio for the two main effects (var00001 and var00002) is presented first.
- For the first variable (sleep deprivation) it is 5.797 which has an exact significance level of 0.017. In other words, this *F*-ratio is statistically significant at the 0.05 level which means there is a difference between the means of two or more of the three sleep conditions.
- Which of the means differ from the others can be further determined by the use of multiple comparison tests such as the unrelated *t*-test.
- For the second variable (alcohol) the *F*-ratio is 31.683 which is significant at less than the 0.0005 level. Since there are only two conditions for this effect we can conclude that the mean score for alcohol (20.22) is significantly higher than that for no-alcohol (11.33).
- The *F*-ratio for the two-way interaction between the two variables is 2.708. As the exact significance level of this ratio is 0.107 we would conclude that there was no significant interaction.

22.3 Reporting the output in Table 22.2

- We could report the results of the output as follows: 'A two-way unrelated ANOVA showed that significant effects were obtained for sleep deprivation

Table 22.3 **Analysis of Variance summary table**

Source of variation	Sums of squares	Degrees of freedom	Mean square	*F*-ratio	Probability
Sleep deprivation	130.11	2	65.06	5.80	<5%
Alcohol	355.56	1	355.56	31.68	<1%
Sleep deprivation with alcohol	60.78	2	30.39	2.71	not significant
Error	134.67	12	11.22		

($F_{2,12}$ = 5.80, p<0.05) and alcohol ($F_{1,12}$ = 31.68, p<0.001) but not for their interaction ($F_{2,12}$ = 2.71, ns).'.

- It is usual to give an Analysis of Variance summary table. A simple one for Table 22.2 would leave out some of the information which is unnecessary (Table 22.3).

Chapter 23

Multiple comparisons in ANOVA

This chapter tells you how to work out which particular pairs of means are significantly different from each other in the analysis of variance. It is used when you have more than two means.

Knowing precisely where significant differences lie between different conditions of your study is important. The overall trend in the ANOVA may tell you only part of the story. SPSS has a number of 'post hoc' procedures which are, of course, applied after the data are collected, not planned initially. They all do slightly different things. There is a thorough discussion of them in Howell (1987). We will illustrate the use of these multiple comparison procedures using the data in Table 23.1, which were previously discussed in Chapter 20.

Table 23.1 **Data for a study of the effects of hormones**

Group 1 Hormone 1	Group 2 Hormone 2	Group 3 Placebo control
9	4	3
12	2	6
8	5	3

23.1 Multiple comparison tests

Quick summary

Enter data

Statistics

Compare Means

One-Way ANOVA . . .

Dependent variable (e.g. var00002)

Independent or factor variable (e.g. var00001)

Define Range . . .

Minimum group code (e.g. 1)

Maximum:

Maximum group code (e.g. 3)

Continue

Post Hoc . . .

Duncan's multiple range test

Tukey's honestly significant difference

Scheffé

Continue

OK

- Enter the data as described in Chapter 20 (Figure 20.1) or retrieve the data file as described in Chapter 1.
- Select Statistics on the menu bar of the Applications window which produces a drop-down menu (Figure 2.6).
- Select Compare Means from this drop-down menu which opens a second drop-down menu.
- Select One-Way ANOVA . . . which opens the One-Way ANOVA dialog box (Figure 20.2)
- Select 'var00002' and the ▶ button beside Dependent List: which puts 'var00002' in this box.
- Select 'var00001' and the ▶ button beside Factor: which puts 'var00001' in this box.
- Select Define Range . . . which opens the One-Way ANOVA: Define Range dialog box (Figure 20.3).
- Type 1 in the box beside Minimum:.
- Select box beside Maximum: and type 3.

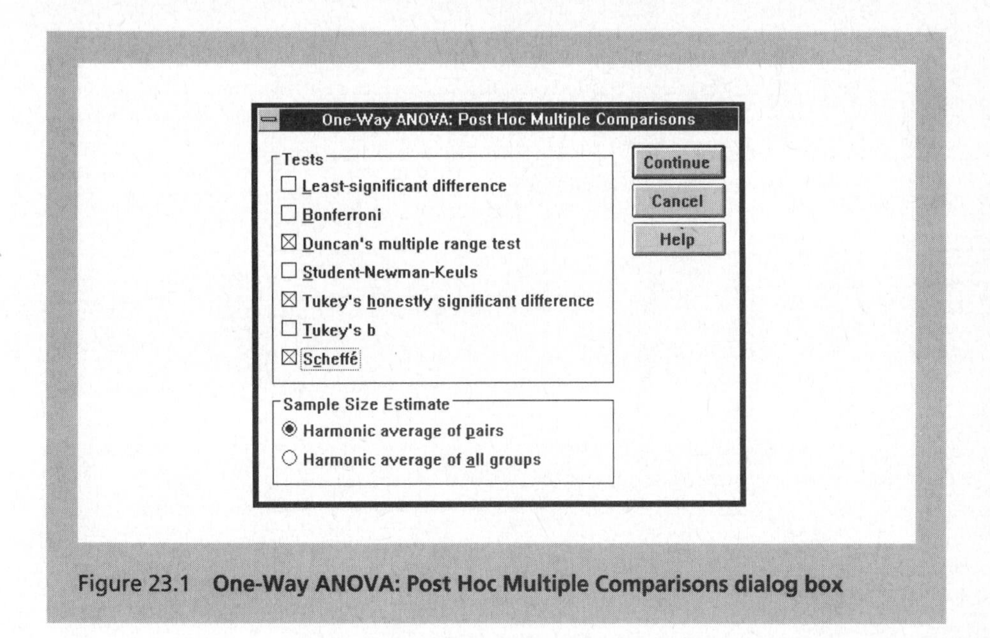

Figure 23.1 **One-Way ANOVA: Post Hoc Multiple Comparisons dialog box**

- Select Continue which closes the One-Way ANOVA: Define Range dialog box.
- Select Post Hoc . . . which opens the One-Way ANOVA: Post Hoc Multiple Comparisons dialog box (Figure 23.1).
- Select Duncan's multiple range test; Tukey's honestly significant difference; and Scheffé .
- Select Continue which closes the One-Way ANOVA: Post Hoc Multiple Comparisons dialog box.
- Select OK which closes the One-Way ANOVA dialog box and the Newdata window and which displays the output shown in Table 23.2.

23.2 Interpreting the output in Table 23.2

- The Analysis of Variance table is presented in the upper part of the output while the results for the three multiple comparison tests are displayed in the lower part of the output.
- The *F*-ratio for the Between Groups effect (i.e. the effects of hormones) is 10.5862 which has an exact significance level of 0.0108. In other words, the Between Groups effect is significant.
- The three multiple comparison tests all suggest the same thing: that there are significant differences between Hormone 1 and Hormone 2, and between Hormone 1 and the Placebo control. There are no other significant differences.

Table 23.2 **Multiple comparison tests**

```
-----ONE  WAY-----

    Variable VAR00002
  By Variable VAR00001

                                    Analysis of
                                     Variance

                                 Sum of         Mean           F          F
Source                 D.F.     Squares       Squares        Ratio      Prob.

Between Groups           2      68.2222       34.1111       10.5862     .0108
Within Groups            6      19.3333        3.2222
Total                    8      87.5556

                          -----ONE  WAY-----

    Variable VAR00002
  By Variable VAR00001

Multiple Range Tests: Duncan test with significance level .05

The difference between two means is significant if
  MEAN(J)-MEAN(I) >= 1.2693 * RANGE * SQRT(1/N(I) + 1/N(J))
  with the following value(s) for RANGE:

  Step          2       3
  RANGE       3.46    3.59

  (*) Indicates significant differences which are shown in the
lower triangle

                         G  G  G
                         r  r  r
                         p  p  p
                         2  3  1
  Mean      VAR00001

  3.6667    Grp 2
  4.0000    Grp 3
  9.6667    Grp 1        *  *

Homogeneous Subsets (highest and lowest means are not signifi-
cantly different)

Subset 1

Group         Grp 2      Grp 3

Mean         3.6667     4.0000

------------------------

Subset 2

Group         Grp 1

Mean         9.6667

------------------------
```

```
-----ONE  WAY-----

    Variable VAR00002
  By Variable VAR00001

Multiple Range Tests: Tukey-HSD test with significance level .050
The difference between two means is significant if
  MEAN(J)-MEAN(I) >= 1.2693 * RANGE * SQRT(1/N(I) + 1/N(J))
  with the following value(s) for RANGE: 4.34

  (*) Indicates significant differences which are shown in the
lower triangle

                         G  G  G
                         r  r  r
                         p  p  p
                         2  3  1

   Mean    VAR00001

  3.6667   Grp 2
  4.0000   Grp 3
  9.6667   Grp 1         *  *

Homogeneous Subsets (highest and lowest means are not signifi-
cantly different)

Subset 1

Group        Grp 2     Grp 3
Mean        3.6667    4.0000
-------------------------

Subset 2
Group        Grp 1
Mean        9.6667
-------------------------

-----ONE  WAY-----

    Variable VAR00002
  By Variable VAR00001

Multiple Range Tests: Scheffe test with significance level .05

The difference between two means is significant if
  MEAN(J)-MEAN(I) >= 1.2693 * RANGE * SQRT(1/N(I) + 1/N(J))
  with the following value(s) for RANGE: 4.54

  (*) Indicates significant differences which are shown in the
lower triangle

                         G  G  G
                         r  r  r
                         p  p  p
                         2  3  1
   Mean    VAR00001
  3.6667   Grp 2
  4.0000   Grp 3
  9.6667   Grp 1         *  *

Homogeneous Subsets (highest and lowest means are not signifi-
cantly different)
```

```
Subset 1

Group        Grp 2     Grp 3
Mean        3.6667    4.0000
-------------------------

Subset 2
Group        Grp 1
Mean        9.6667
```

So for example, it is not possible to say that Hormone 2 and the Placebo control are significantly different.

- The choice between the three tests is not a simple matter. Howells (1987) makes some recommendations.

23.3 Reporting the output in Table 23.2

We could report the results of the output as follows: 'A one-way unrelated Analysis of Variance showed an overall significant effect for the type of drug treatment ($F_{2,6} = 10.59$, $p = 0.01$). Scheffé's range test found thatthe Hormone 1 group differed from the remaining two groups at the 5% level but no other significant differences were found.'.

23.4 Reference

Howell, D. (1987). *Statistical Methods for Psychology*. Boston: Duxbury Press.

Chapter 24

Analysis of covariance (ANCOVA)

The analysis of covariance allows you to control or adjust for variables which correlate with your dependent variable before comparing the means on the dependent variable.

24.1 Introduction

One of the computations which are complex by hand but quick and easily done on SPSS is the analysis of covariance. This is much the same as the analysis of variance dealt with elsewhere but with one major difference. This is that the effects of additional variables (covariates) are taken away as part of the analysis. It is a bit like using partial correlation to get rid of the effects of a third variable on a correlation. We will illustrate the computation of an analysis of covariance (ANCOVA) with the data shown in Table 24.1 which are the same as those presented in Table 20.1 except that depression scores taken immediately prior to the three treatments have been included.

It could be that differences in depression prior to the treatment affect the outcome of the analysis. Essentially by adjusting the scores on the dependent variable to 'get rid' of these pre-existing differences, it is possible to disregard the possibility that these pre-existing differences are affecting the analysis. So, if (a) the pre-treatment or test scores are correlated with the post-treatment or test scores, and (b) the pre-test scores differ between the three treatments, then these pre-test differences can be statistically controlled by covarying them out of the analysis.

Table 24.1 **Data for a study of the effects of hormones (analysis of covariance)**

Group 1 Hormone 1		Group 2 Hormone 2		Group 3 Placebo control	
Pre	**Post**	**Pre**	**Post**	**Pre**	**Post**
5	9	3	4	2	3
4	12	2	2	3	6
6	8	1	5	2	3

24.2 One-way ANCOVA

Quick summary

Enter data

Statistics

ANOVA Models

Simple Factorial . . .

Dependent variable and ▶ button beside Dependent:

Independent variable and ▶ button beside Factor[s]:

Define Range . . .

Minimum group code (e.g. 1)

Maximum:

Maximum group code (e.g. 3)

Covariate and ▶ button beside Covariate[s]:

Options . . .

Hierarchical or Experimental

Means and counts

MCA

Continue

OK

- Enter the data in Table 24.1 in Newdata as shown in Figure 24.1. The pre-test depression scores have been put into the third column.
- Select Statistics on the menu bar of the Applications window which produces a drop-down menu (Figure 2.6).
- Select ANOVA Models from this drop-down menu which opens a second drop-down menu.
- Select Simple Factorial . . . which opens the Simple Factorial ANOVA dialog box (Figure 24.2).
- Select 'var00002' and the ▶ button beside Dependent: which puts 'var00002' in this box.

	var00001	var00002	var00003	var
1	1.00	9.00	5.00	
2	1.00	12.00	4.00	
3	1.00	8.00	6.00	
4	2.00	4.00	3.00	
5	2.00	2.00	2.00	
6	2.00	5.00	1.00	
7	3.00	3.00	2.00	
8	3.00	6.00	3.00	
9	3.00	3.00	2.00	
10				

Figure 24.1 **Post- and pre-test depression scores in three treatments**

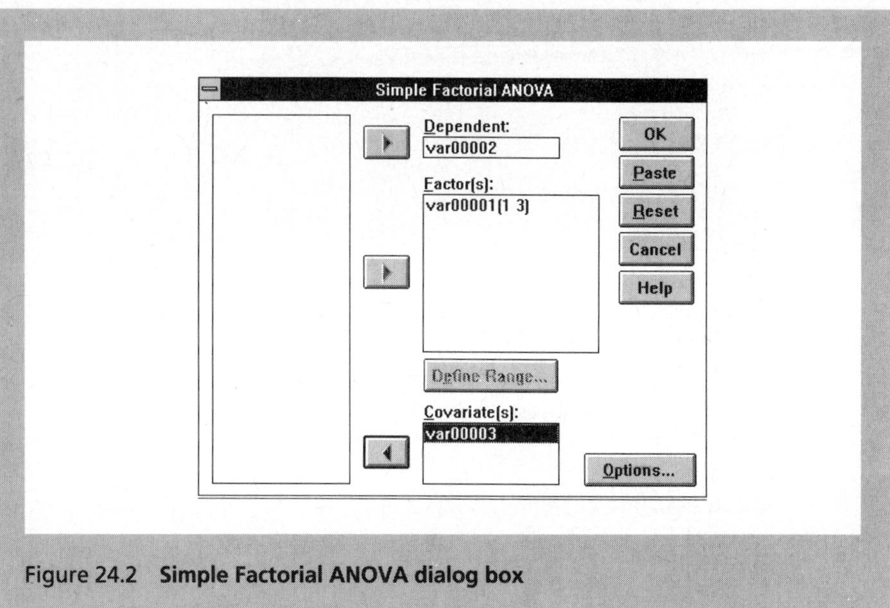

Figure 24.2 **Simple Factorial ANOVA dialog box**

- Select 'var00001' and the ▶ button beside Factor[s]: which puts 'var00001' in this box.
- Select Define Range . . . which opens the Simple Factorial ANOVA: Define Range dialog box (Figure 22.3).
- Type 1 in the box beside Minimum:.

Table 24.2 **One-way ANCOVA output**

```
                    * * * C E L L   M E A N S * * *
        VAR00002
     by VAR00001
Total Population
  5.78
  (  9)
VAR00001
     1          2          3
  9.67       3.67       4.00
  (  3)      (  3)      (  3)
           * * * A N A L Y S I S   O F   V A R I A N C E * * *
        VAR00002
  by    VAR00001
  with  VAR00003
     HIERARCHICAL sums of squares
     Covariates entered FIRST
                              Sum of              Mean                Sig
Source of Variation          Squares     DF      Square         F     of F
Covariates                    43.726      1      43.726    12.561     .016
  VAR00003                    43.726      1      43.726    12.561     .016
Main Effects                  26.425      2      13.213     3.796     .099
  VAR00001                    26.425      2      13.213     3.796     .099
Explained                     70.151      3      23.384     6.718     .033
Residual                      17.405      5       3.481
Total                         87.556      8      10.944
9 cases were processed.
0 cases ( .0 pct) were missing.
  * * * M U L T I P L E   C L A S S I F I C A T I O N   A N A L Y S I S * * *
           VAR00002
     by    VAR00001
     with  VAR00003
Grand Mean = 5.78                                      Adjusted for
                                                       Independents
                                     Unadjusted        + Covariates
Variable + Category           N      Dev'n   Eta     Dev'n      Beta
VAR00001
    1                         3       3.89            5.10
    2                         3      -2.11           -2.83
    3                         3      -1.78           -2.28
                                              .88                1.16
Multiple R Squared                                                .801
Multiple R                                                        .895
```

- Select box beside Maximum: and type 3.
- Select Continue which closes the Simple Factorial ANOVA: Define Range dialog box.
- Select 'var00003' and the ▶ button beside Covariate[s]: which puts 'var00003' in this box.
- Select Options . . . which opens the Simple Factorial ANOVA: Options dialog box (Figure 22.4).
- Select either Hierarchical or Experimental in the Method section.
- Select Means and counts in the Statistics section.
- Select MCA.
- Select Continue which closes the Simple Factorial ANOVA: Options dialog box.
- Select OK which closes the Simple Factorial ANOVA dialog box and the Newdata window and which displays the output shown in Table 24.2.

24.3 Interpreting the output in Table 24.2

- The unadjusted means are displayed in the upper part of the output, the Analysis of Covariance table is presented in the middle part of the output and the multiple classification table in the lower part of the table.
- The multiple classification table is used to find the adjusted means of the three treatments. These are the means when all groups are adjusted to be identical on the covariate (in this case pre-treatment depression scores).
- To obtain the adjusted means we add the deviation (Dev'n) for each of the three treatments Adjusted for Independents + Covariates to the Grand Mean. The adjusted mean is 10.88 (5.78 + 5.10 = 10.88) for the first treatment, 2.95 (5.78 + (–2.83) = 2.95) for the second treatment and 3.50 (5.78 + (–2.28) = 3.50) for the third treatment.
- We can see that these adjusted means seem to differ from the unadjusted means shown in the first part of the output. For the first treatment the adjusted mean is 10.88 and the unadjusted mean is 9.67. For the second treatment the adjusted mean is 2.95 and the unadjusted mean is 3.67, while for the third treatment the adjusted mean is 3.50 and the unadjusted mean is 4.00.
- The *F*-ratio for the main effect is 3.796 (13.213/3.481 = 3.796).
- The probability of this *F*-ratio is 0.099. In other words, it is greater than the 0.05 critical value and so is not statistically significant.

24.4 Reporting the output in Table 24.2

- We could report the results of the output as follows: 'A one-way ANCOVA showed that when pre-test depression was covaried out, the main effect of

Table 24.3 **ANCOVA summary table for effects of treatments on depression controlling for pre-treatment depression**

Source of variance	**Sums of squares**	**Degrees of freedom**	**Mean square**	***F*-ratio**
Covariate (pre-treatment depression scores)	43.73	1	43.73	12.56*
Main effect (treatment)	26.43	2	13.21	3.80
Total explained	70.15	3	23.38	6.72*
Residual error	17.41	5	3.48	

* Significant at 5% level.

treatment onpost-test depression was not significant ($F_{2,5}$ = 3.79, ns). The covariate, pre-treatment depression scores, had a significant effect on post-treatment depression scores.'. You would normally also report the changes to the means once the covariate has been removed.

- In addition, we would normally give an ANCOVA summary table as in Table 24.3. Notice that the Total explained is the sum of the covariate and main effect sums of squares.

Chapter 25

Reading ASCII text files into Newdata

Sometimes you have a computer file of data which you wish to use on SPSS. This chapter tells you how to use data not specifically entered into the SPSS Newdata spreadsheet.

25.1 Introduction

SPSS for Windows is obviously one of many different computer programs for analysing data. There are circumstances in which researchers might wish to take data sets which have been prepared for another computer program and run those data through SPSS for Windows. It can be expensive in time and/or money to re-enter data, say, from a big survey into the Newdata spreadsheet. Sometimes, the only form in which the data are available is as an archive electronic data file; the original questionnaires may have been thrown away. No matter the reason for using an imported data file, SPSS for Windows can accept files in other forms. In particular, data files are sometimes written as simple text or ASCII files as these can be readily transferred from one type of computer to another. ASCII stands for American Standard Code for Information Interchange. To analyse an ASCII data file you first need to read it into Newdata.

Suppose, for example, that you had an ASCII data file called data.txt which consisted of the following numbers:

```
1118
2119
3218
```

Obviously you cannot sensibly use an ASCII file until you know exactly where the information for each variable is. However, we do know where and what the information is for the above small file. The figures in the first column simply number the three different participants for whom we have data. The values in the second column contain the code for gender with 1 representing females and 2 males, while the values in the third and fourth columns indicate the age of the three people. We would carry out the following procedure to enter this ASCII data file into Newdata.

25.2 Reading an ASCII data file

Quick summary

File

Read ASCII Data . . .

Disk drive, directory and file name of ASCII data file

SPSS variable

Start Column number

End Column number

Add

Repeat as necessary

OK

- Select File from the menu bar in the Applications window to produce a drop-down menu (Figure 1.5).
- Select Read ASCII Data . . . which opens the Read ASCII Data File dialog box (Figure 25.1).

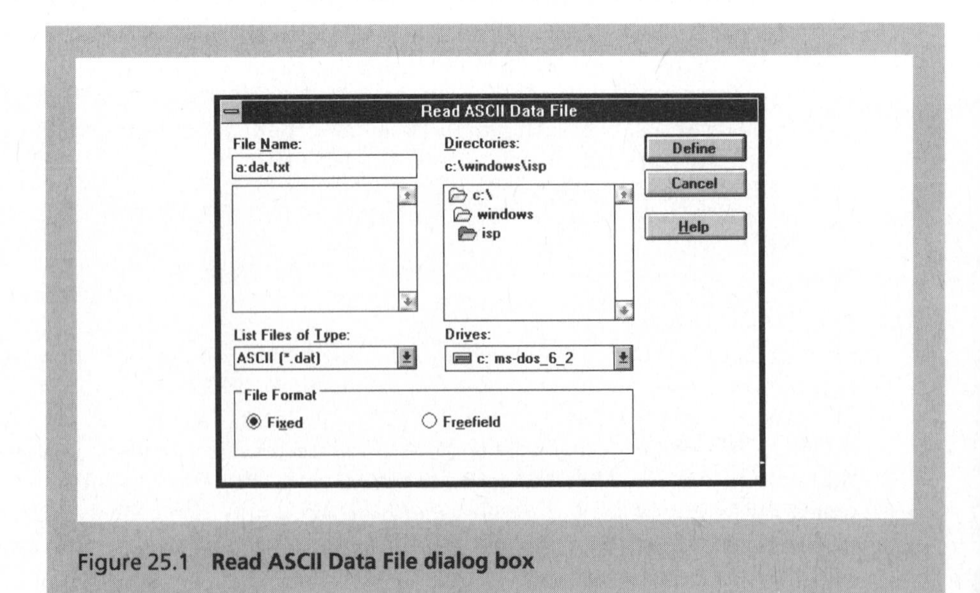

Figure 25.1 **Read ASCII Data File dialog box**

Define Fixed Variables
File: A:\DAT.TXT
Name: gender
Record: 1
Start Column: 2
End Column:
Data Type
123=123, 1.23=1.23
Numeric as is
Defined Variables:
OK
Paste
Reset
Cancel
Help
Add
Change
Remove
Value Assigned to Blanks for Numeric Variables
System missing
Value:
Display summary table
Display warning message for undefined data

Figure 25.2 **Define Fixed Variables dialog box**

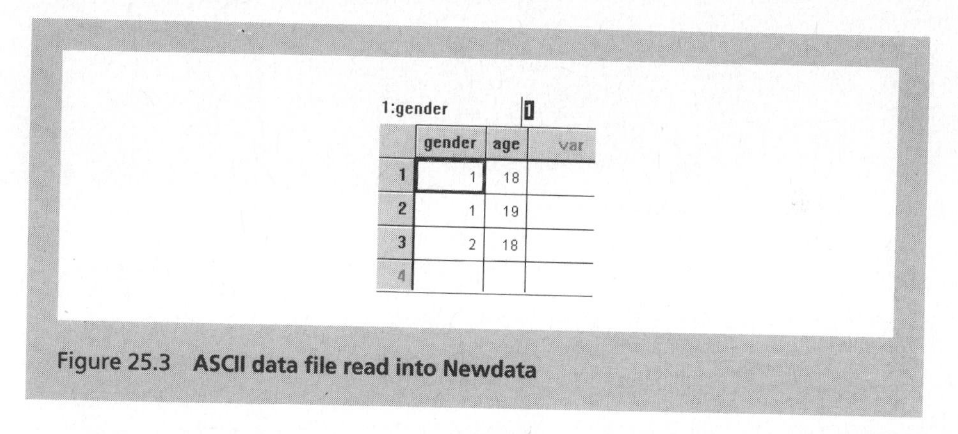

1:gender 1

	gender	age	var
1	1	18	
2	1	19	
3	2	18	
4			

Figure 25.3 **ASCII data file read into Newdata**

- Type in the disk drive, directory (if any) and file name of the ASCII data file (e.g. a:dat.txt) in the File Name: box.
- Select Define which opens the Define Fixed Variables dialog box (Figure 25.2).
- Type in the Name: box the SPSS name of the first variable to be defined. As the rows of Newdata are already numbered, the name of the first variable to be defined is gender.
- The column in which gender is stored is column 2. As there is only one column for gender, the number to be put in the Start Column: box is the same

as the number for the End Column: box. Consequently, it is only necessary to put 2 in the Start Column: box. Select the Start Column: box and type 2.

- Select Add which puts 1 2- 2 gender in the Defined Variables: box.
- Type in the Name: box the SPSS name of the next variable to be defined which is age.
- Select the Start Column: box and type 3.
- Select the End Column: box and type 4.
- Select Add which puts 1 3- 4 age in the Defined Variables: box.
- After all the required variables have been defined select OK which closes the Define Fixed Variables dialog box and opens the Newdata window which now contains the SPSS variable names and values of the ASCII data file as shown in Figure 25.3.

Chapter 26

Partial correlation

If you suspect that a correlation between two variables is affected by their correlations with yet another variable, it is possible to adjust for the effects of this additional variable using the partial correlation procedure.

SPSS for Windows cannot easily compute partial correlations from a matrix of zero-order correlations. Consequently, we will illustrate the computation of partial correlations with the raw scores in Table 26.1 which represent a numerical intelligence test score, a verbal intelligence test score and age in years. We will correlate the two test scores partialling out age.

Table 26.1 **Numerical and verbal intelligence test scores and age**

Numerical scores	Verbal scores	Age
90	90	13
100	95	15
95	95	15
105	105	16
100	100	17

26.1 Partial correlation

Quick summary

Statistics

Correlate

Partial . . .

Variables and ▶ *Variables:*

Control variables and ▶ *Controlling for:*

OK

- Enter the numerical intelligence test scores in the first column of Newdata, the verbal intelligence test scores in the second column and age in the third column (Figure 26.1).
- Select Statistics from the menu bar in the Applications window which produces a drop-down menu (Figure 2.6).

	var00001	var00002	var00003	var
1	90.00	90.00	13.00	
2	100.00	95.00	15.00	
3	95.00	95.00	15.00	
4	105.00	105.00	16.00	
5	100.00	100.00	17.00	
6				

Figure 26.1 **Numerical and verbal intelligence scores and age in Newdata**

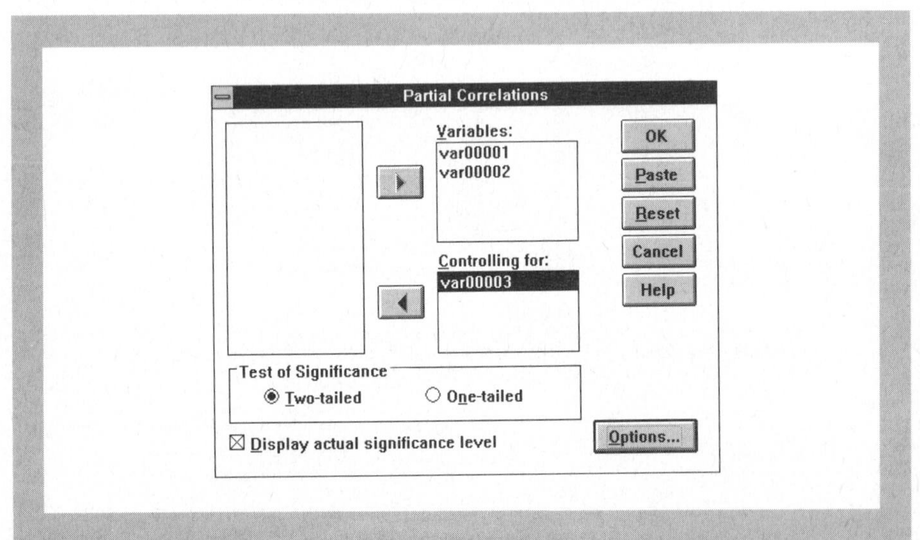

Figure 26.2 **Partial Correlations dialog box**

- Select Correlate from the drop-down menu which reveals a smaller drop-down menu.
- Select Partial . . . from this drop-down menu which opens the Partial Correlations dialog box (Figure 26.2).
- Select 'var00001' and 'var00002' and the ▶ button beside the Variables: text box which puts 'var00001' and 'var00002' in it.
- Select 'var00003' and the ▶ button beside the Controlling for: text box which puts 'var00003' in it. (You can include several other control variables at this point if they are available. For our example, however, we only have one.)
- Select OK which closes the Partial Correlations dialog box and the Newdata window and which displays the output in Table 26.2 in the Output window.

Table 26.2 **Partial correlation output**

```
--- PARTIAL CORRELATION COEFFICIENTS ---

Controlling for.. VAR00003

               VAR00001    VAR00002

VAR00001        1.0000       .7762
               (    0)     (    2)
               P= .        P= .224

VAR00002         .7762      1.0000
               (    2)     (    0)
               P= .224     P= .

(Coefficient / (D.F.) / 2-tailed Significance)

'' . '' is printed if a coefficient cannot be computed
```

26.2 Interpreting the output in Table 26.2

- The variables on which the partial correlation was carried out are given both in the columns and in the rows. We have just two variables so a 2 × 2 correlation matrix is generated.
- The print-out gives indications of how to read the entries in the table (`Coefficient / (D.F.) / 2-tailed Significance`)
- The partial correlation (`Coefficient`) of VAR00001 with VAR00002 controlling for VAR00003 is 0.7762.
- The degrees of freedom (`D.F.`) are 2.
- The exact significance level (`2-tailed Significance`) is given to three decimal places (`P= .224`).

- Partial correlations are displayed in a matrix. The diagonal of this matrix (from top left to bottom right) consists of the variable correlated with itself which obviously gives a perfect correlation of 1.0000. No significance level is given for this value as it never varies (`P=   .`).
- The values of the partial correlations are symmetrical around the diagonal from top right to bottom left in the matrix.

26.3 Reporting the output in Table 26.2

If you calculate the correlation between numerical intelligence and verbal intelligence the Pearson correlation is 0.92. Bearing this in mind, we could report the results in Table 26.2 as follows: 'The correlation between numerical intelligence and verbal intelligence is 0.92 (significant at the 5% level with a two-tailed test). However, the correlation between numerical intelligence and verbal intelligence controlling for age declines to 0.78 (not significant). In other words, there is no significant relationship between numerical and verbal intelligence when age is controlled for.'.

Chapter 27

Factor analysis

Factor analysis allows you to make sense of a complex set of variables by reducing them to a smaller number of factors (or supervariables) which account for many of the original variables.

We will illustrate the computation of a principal axis factor analysis with the data shown in Table 27.1 which consist of scores on six variables for nine individuals. This is only for illustrative purposes; it would be considered a ludicrously small number of cases to do a factor analysis on. Normally, you should think of having 10 or 15 times as many cases as you have variables. The following is a standard factor analysis which is adequate for most situations. However, SPSS has many options for factor analysis.

Table 27.1 **Scores of nine individuals on six variables**

Individual	Variable 1	Variable 2	Variable 3	Variable 4	Variable 5	Variable 6
1	10	15	8	26	15	8
2	6	16	5	25	12	9
3	2	11	1	22	7	6
4	5	16	3	28	11	9
5	7	15	4	24	12	7
6	8	13	4	23	14	6
7	6	17	3	29	10	9
8	2	18	1	28	8	8
9	5	14	2	25	10	6

27.1 Principal axis analysis with rotation

Quick summary

Enter data

Statistics

Data Reduction

Factor . . .

Variables ▶

Extraction . . .

Scree plot

Principal components

Principal-axis factoring

Continue

Descriptives . . .

Coefficients

Continue

Rotation . . .

Varimax

Continue

Options . . .

Sorted by size

Continue

OK

---

	var00001	var00002	var00003	var00004	var00005	var00006	var
1	10.00	15.00	8.00	26.00	15.00	8.00	
2	6.00	16.00	5.00	25.00	12.00	9.00	
3	2.00	11.00	1.00	22.00	7.00	6.00	
4	5.00	16.00	3.00	28.00	11.00	9.00	
5	7.00	15.00	4.00	24.00	12.00	7.00	
6	8.00	13.00	4.00	23.00	14.00	6.00	
7	6.00	17.00	3.00	29.00	10.00	9.00	
8	2.00	18.00	1.00	28.00	8.00	8.00	
9	5.00	14.00	2.00	25.00	10.00	6.00	
10							

Figure 27.1 **Six scores in Newdata**

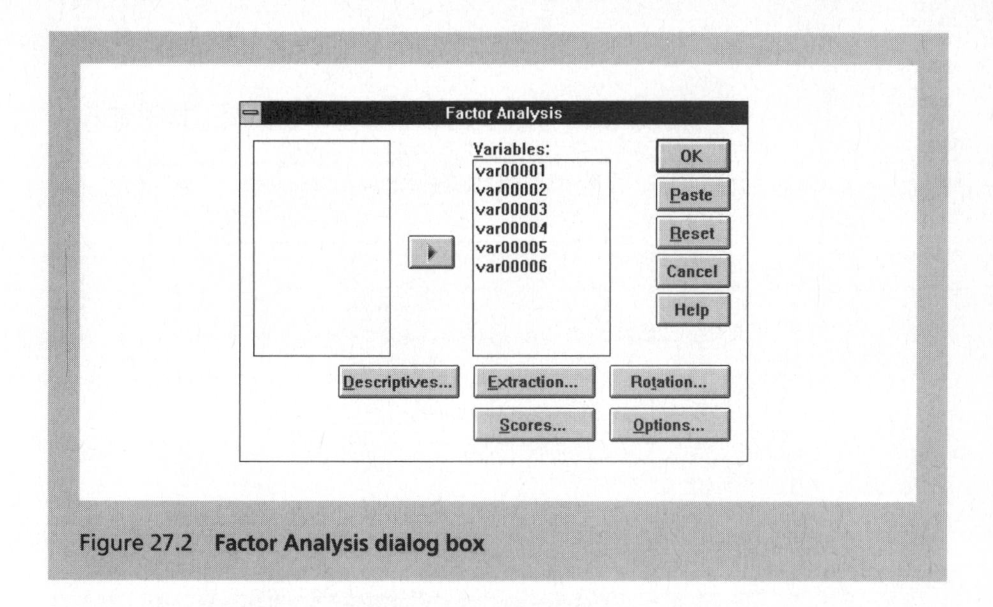

Figure 27.2 **Factor Analysis dialog box**

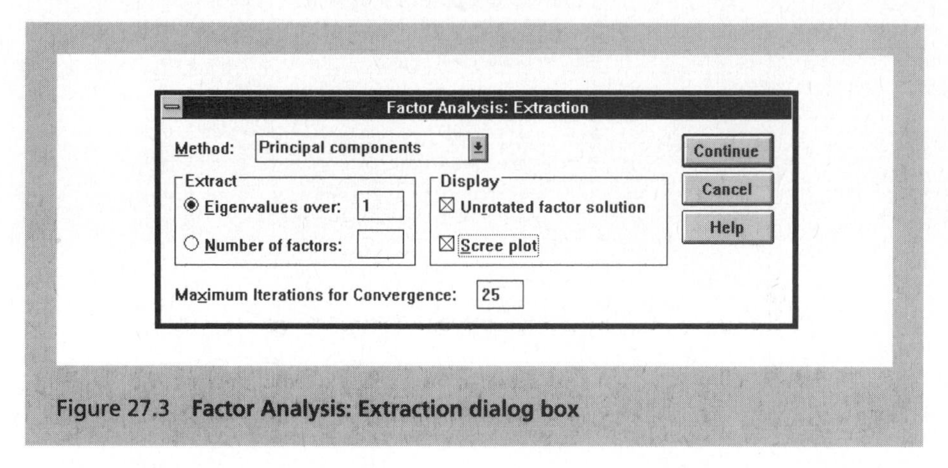

Figure 27.3 **Factor Analysis: Extraction dialog box**

- Enter the data of Table 27.1 into Newdata as shown in Figure 27.1.
- Select Statistics from the menu bar in the Applications window which produces a drop-down menu (Figure 2.6).
- Select Data Reduction from the drop-down menu which reveals a smaller drop-down menu.
- Select Factor . . . from this drop-down menu which opens the Factor Analysis dialog box (Figure 27.2).
- Select 'var00001' to 'var00006' and then the ▶ button which puts these six variables in the Variables: text box.

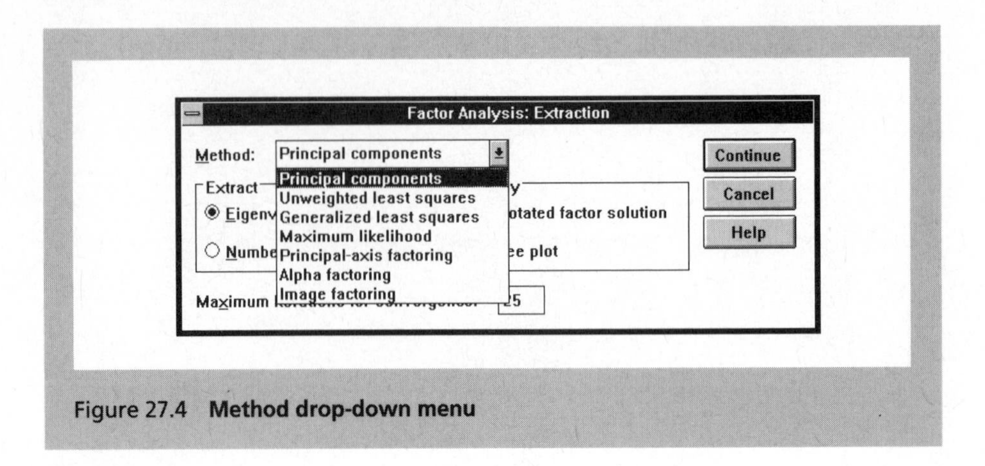

Figure 27.4 **Method drop-down menu**

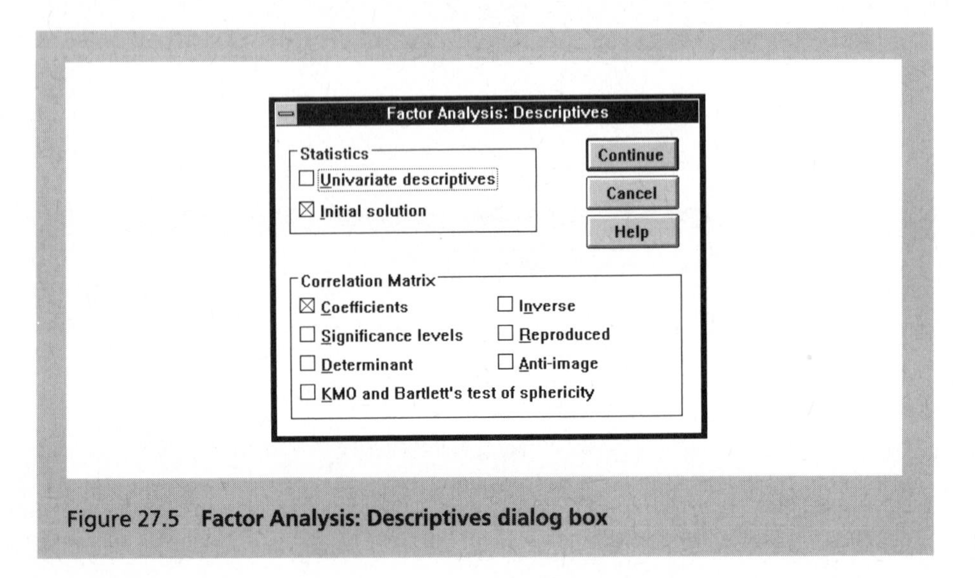

Figure 27.5 **Factor Analysis: Descriptives dialog box**

- Select Extraction . . . which opens the Factor Analysis: Extraction dialog box (Figure 27.3).
- Select Scree plot.
- Select Principal components or the down-pointing arrow in the Method: box which produces a drop-down menu (Figure 27.4).
- Select Principal-axis factoring which puts this term in the Method: box.
- Select Continue which closes the Factor Analysis: Extraction dialog box.
- Select Descriptives . . . which opens the Factor Analysis: Descriptives dialog box (Figure 27.5).

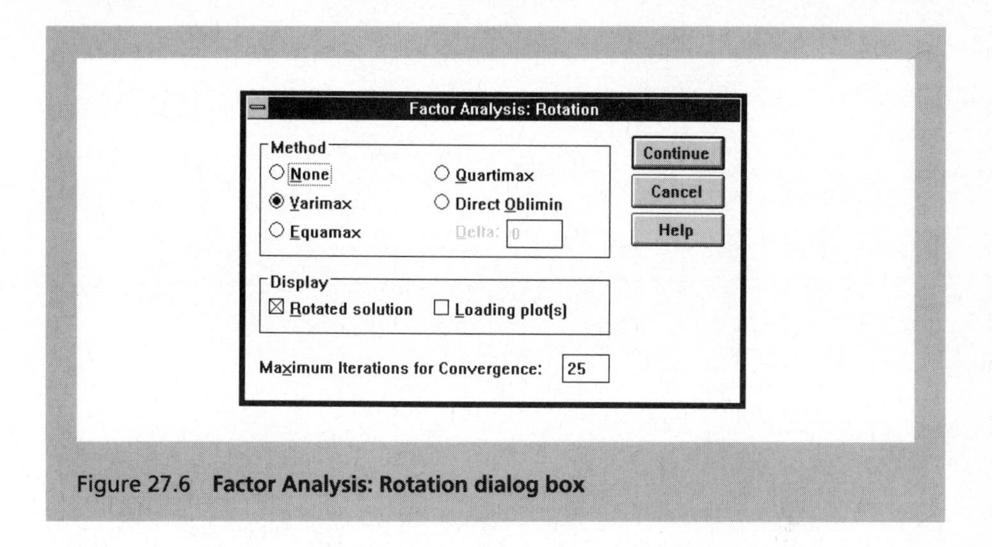

Figure 27.6 **Factor Analysis: Rotation dialog box**

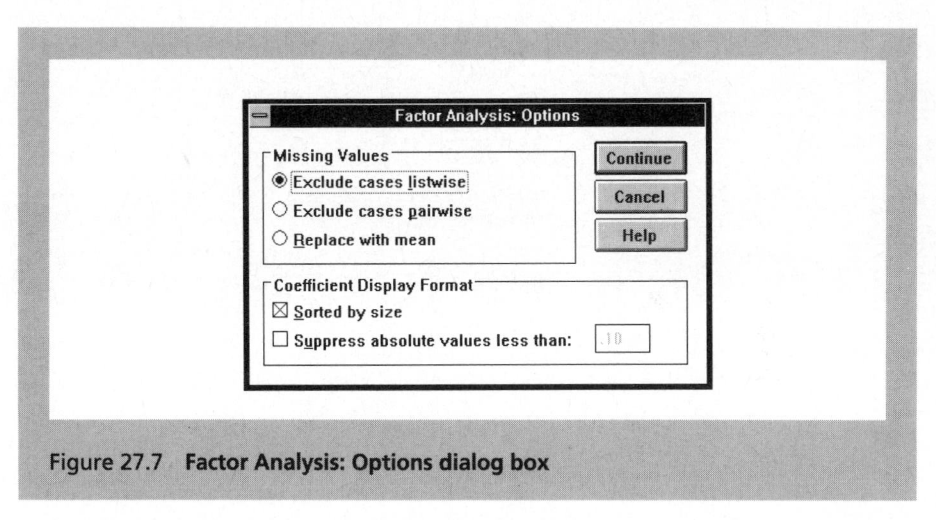

Figure 27.7 **Factor Analysis: Options dialog box**

- Select Coefficients in the Correlation Matrix section to obtain a correlation matrix of the six variables.
- Select Continue to close the Factor Analysis: Descriptives dialog box.
- Select Rotation . . . to open the Factor Analysis: Rotation dialog box (Figure 27.6).
- Select Varimax in the Method section to obtain orthogonally rotated factors.
- Select Continue to close the Factor Analysis: Rotation dialog box.
- Select Options . . . to open the Factor Analysis: Options dialog box (Figure 27.7).

Table 27.2 **Principal axis analysis output**

```
--------------FACTOR ANALYSIS---------------

Analysis number 1  Listwise deletion of cases with missing
values

Correlation Matrix:

         VAR00001 VAR00002 VAR00003 VAR00004 VAR00005 VAR00006

VAR00001 1.00000
VAR00002   .00000  1.00000
VAR00003   .90980   .08087  1.00000
VAR00004 -.04670   .88252  -.00529  1.00000
VAR00005   .96296   .02268   .90246  -.08006  1.00000
VAR00006   .09623   .79550   .29070   .78871   .10825  1.00000

Extraction  1 for analysis  1, Principal Axis Factoring (PAF)

Initial Statistics:

Variable    Communality  *  Factor  Eigenvalue  Pct of Var  Cum Pct
                         *
VAR00001        .95672   *    1      2.95114       49.2      49.2
VAR00002        .83219   *    2      2.57885       43.0      92.2
VAR00003        .91347   *    3       .26405        4.4      96.6
VAR00004        .86996   *    4       .12373        2.1      98.6
VAR00005        .94342   *    5       .05844        1.0      99.6
VAR00006        .82656   *    6       .02378         .4     100.0

-------------FACTOR ANALYSIS-------------

Hi-Res Chart  # 1:Factor scree plot

PAF  extracted  2 factors.  8 iterations required.

Factor Matrix:

          Factor 1  Factor 2

VAR00003   .91072   -.24236
VAR00001   .90650   -.36348
VAR00005   .90379   -.36247
VAR00004   .27224    .90041
VAR00002   .34621    .86634
VAR00006   .44643    .74537

Final Statistics:

Variable    Communality  *  Factor  Eigenvalue  Pct of Var  Cum Pct
                         *
VAR00001       .95387    *    1      2.86127       47.7      47.7
VAR00002       .87041    *    2      2.43911       40.7      88.3
```

```
VAR00003     .88815     *
VAR00004     .88485     *
VAR00005     .94822     *
VAR00006     .75488     *

VARIMAX  rotation  1 for extraction  1 in analysis 1 – Kaiser
Normalization.

VARIMAX converged in 3 iterations.

Rotated Factor Matrix:

            Factor 1  Factor 2
VAR00001    .97661    –.00941
VAR00005    .97372    –.00946
VAR00003    .93656     .10496

--------------F A C T O R   A N A L Y S I S-------------

            Factor 1  Factor 2
VAR00004   –.07338    .93780
VAR00002    .00791    .93293
VAR00006    .14523    .85661

Factor Transformation Matrix:

            Factor 1  Factor 2
Factor 1    .93171    .36320
Factor 2   –.36320    .93171
```

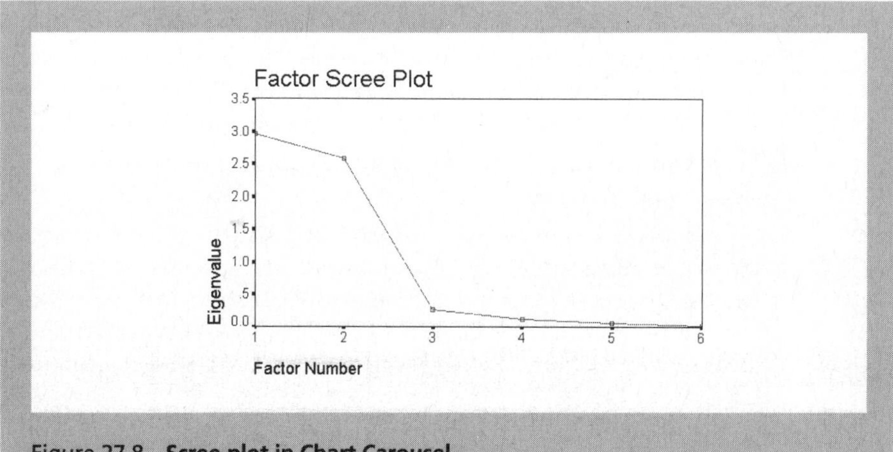

Figure 27.8 **Scree plot in Chart Carousel**

- Select Sorted by size in the Coefficient Display Format section to sort the factor loadings by size.
- Select Continue to close the Factor Analysis: Options dialog box.
- Select OK which closes the Factor Analysis dialog box and the Newdata window and which displays the output in Table 27.2 in the Output window.
- To display the scree plot shown in Figure 27.8 select Window which unfolds a drop-down menu.
- Select 3 Chart Carousel.

27.2 Interpreting the output in Table 27.2

- The correlation matrix is presented first. From this it appears that there are two groups of variables that are strongly intercorrelated. One consists of VAR00001, VAR00003 and VAR00005 and the other of VAR00002, VAR00004 and VAR00006. Normally in factor analysis the correlation matrix is much more difficult to decipher than this. Our data are highly stylised.
- **Two principal axis factors were initially extracted in this case. The computer ignores factors with an eigenvalue of less than 1.00. This is because such factors consist of uninterpretable error variation. Of course, your analysis may have even more (or fewer) factors.**
- The Scree test also shows that a break in the size of eigenvalues for the factors occurs after the second factor: the curve is fairly flat after the second factor. Since it is important in factor analysis to ensure that you do not have too many factors, you may wish to do your factor analysis and rotation stipulating the number of factors once you have the results of the Scree test. (This can be done by inserting the Number of factors in the Factor Analysis Extraction dialog box.) In the case of our data this does not need to be done since the computer has used the first two factors and ignored the others because of the minimum eigenvalue requirement of 1.00. It is not unusual for a factor analysis to be recomputed in the light of the pattern which emerges.
- **These two factors are then orthogonally rotated and the loadings of the six variables on these two factors are shown in the Rotated Factor Matrix.**
- **The variables are ordered or sorted according to their loading on the first factor. Variables which are highly loaded are grouped together in order of their loading on the first factor. Those which load poorly on the first factor are grouped together at the bottom of the list. This helps interpretation of the factor since the high loading items are the ones which primarily help you decide what the factor might be.**
- On the first factor, VAR00001 has the highest loading (0.97661) followed by VAR00005 (0.97372) and VAR00003 (0.93656).

- On the second factor, VAR00004 has the highest loading (0.93780) followed by VAR00002 (0.93293) and VAR00006 (0.85661).
- We would interpret the meaning of these factors in terms of the content of the variables that loaded most highly on them.
- The percentage of variance that each of the orthogonally rotated factors accounts for is not given but can be easily worked out: square the loading of the variables on each factor; sum the squared loadings; divide by the number of variables; and multiply the result by 100. So the percentage of variance accounted for by the first factor is 47 and by the second factor is 42.

27.3 Reporting the output in Table 27.2

- It would be usual to tabulate the factors and variables, space permitting. Since the data in our example are on various tests of skill, the factor analysis table might be as in Table 27.3. The figures have been given to two decimal places.
- The exact way of reporting the results of a factor analysis will depend on the purpose of the analysis. One way of describing the results would be as follows: 'A principal axis factor analysis was conducted on the correlations of the six variables. Two factors were initially extracted with eigenvalues equal to or greater than 1.00. Orthogonal rotation of the factors yielded the factor structure given in Table 27.3. The first factor accounted for 47% of the variance and the second factor 42%. The first factor seems to be hand–eye coordination and the second factor seems to be verbal flexibility.'. With factor analysis, since the factors have to be interpreted, differences in interpretation may occur.

Table 27.3 **Orthogonal factor loading matrix for six skills**

Variable	Factor 1	Factor 2
Skill at batting	0.98	–0.01
Skill at crosswords	0.01	0.93
Skill at darts	0.94	0.10
Skill at 'Scrabble'	–0.07	0.94
Skill at juggling	0.97	–0.01
Skill at spelling	0.15	0.86

Chapter 28

Stepwise multiple regression

Stepwise multiple regression is a way of choosing predictors of a particular dependent variable on the basis of statistical criteria. Essentially the statistical procedure decides which independent variable is the best predictor, the second best predictor, etc.

28.1 Introduction

We will illustrate the computation of a stepwise multiple regression analysis with the data shown in Table 28.1 which consist of scores for six individuals on the four variables of educational achievement, intellectual ability, school motivation and parental interest respectively.

Because this is for illustrative purposes and to save space, we are going to enter these data 20 times to give us a respectable amount of data to work with. Obviously you would not do this if your data were real. It is important to use quite a lot of research participants or cases for multiple regression. Ten or 15 times your number of variables would be reasonably generous. Of course, you can use fewer for data exploration purposes.

Table 28.1 **Data for stepwise multiple regression**

Educational achievement	Intellectual ability	School motivation	Parental interest
1	2	1	2
2	2	3	1
2	2	3	3
3	4	3	2
3	3	4	3
4	3	2	2

28.2 Stepwise multiple regression analysis

Quick summary

Enter data

Statistics

Regression

Linear . . .

Criterion and ▶ beside Dependent:

Predictors and ▶ beside Independent[s]:

Enter in Method:

Stepwise

OK

- Enter the data in Table 28.1 in Newdata, putting the scores for educational achievement in the first column, the scores for intellectual ability in the second column, the scores for school motivation in the third column, and the scores for parental interest in the fourth column, as shown in Figure 28.1.
- Carry out the instructions in the box *only if you are following the present particular analysis which enters the data 20 times. Otherwise ignore the box.*

	var00001	var00002	var00003	var00004	var
1	1.00	2.00	1.00	2.00	
2	2.00	2.00	3.00	1.00	
3	2.00	2.00	3.00	3.00	
4	3.00	4.00	3.00	2.00	
5	3.00	3.00	4.00	3.00	
6	4.00	3.00	2.00	2.00	
7					

Figure 28.1 **Scores of four variables in Newdata**

- In this particular analysis, because the size of the sample is too small for any of the predictors to explain a significant proportion of the variance in the criterion, we will increase the size of this sample 20 times by copying the data and pasting it in the first empty row 19 times using the Edit option shown in Figure 28.2.
- To do this move the cursor to the cell in the first column of the first row in Newdata, press the left button of the mouse and, holding it down, move the cursor to the cell in the fourth column of the sixth row. The data are now highlighted.
- Select Edit to produce the drop-down menu shown in Figure 28.2.
- Select Copy which removes the drop-down menu.
- Move the cursor to the cell in the first column of the seventh row.
- Select Edit to produce the drop-down menu again.
- Select Paste which removes the drop-down menu and inserts the copied material in the next six rows of the first four columns.
- Repeat this procedure 18 times. Obviously you could reduce this repetition by copying each new set of data as they are produced, copying 12 rows the next time, 24 rows after that and so on.
- Alternatively it is probably quicker to weight the six rows as described in Chapter 6.
- Now return to the main instructions (select Statistics).

- Select Statistics from the menu bar in the Applications window which produces a drop-down menu (Figure 2.6).
- Select Regression from the drop-down menu which reveals a smaller drop-down menu.
- Select Linear . . . from this drop-down menu which opens the Linear Regression dialog box (Figure 8.1).
- Select 'var00001' and then the ▶ button beside the Dependent: box which puts 'var00001' in this box.
- Select 'var00002', 'var00003' and 'var00004' and then the ▶ button beside the Independent[s]: box which puts these variables in this box.
- Select Enter or the down-pointing arrow in the Method: box to produce a small drop-down menu.
- Select Stepwise which puts Stepwise in the Method: box.
- Select OK which closes the Linear Regression dialog box and the Newdata window and which displays the output shown in Table 28.2 in the Output window.

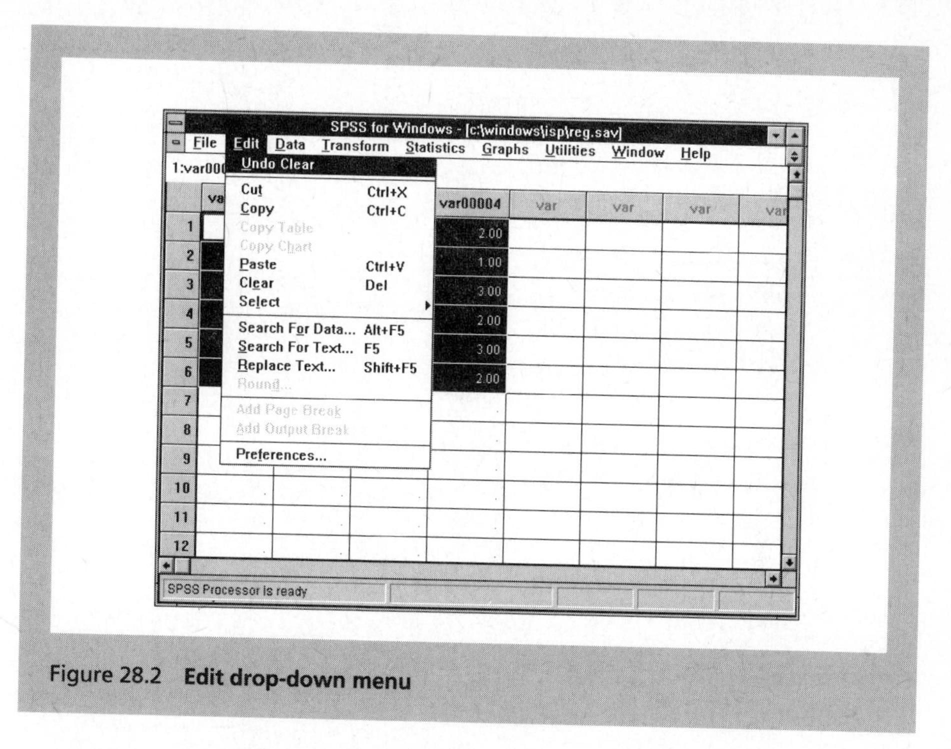

Figure 28.2 **Edit drop-down menu**

Table 28.2 **Stepwise multiple regression analysis output**

```
* * * * MULTIPLE REGRESSION * * * *

Listwise Deletion of Missing Data

Equation Number 1    Dependent Variable..    VAR00001

Block Number 1. Method: Stepwise    Criteria PIN .0500 POUT .1000
   VAR00002   VAR00003   VAR00004

Variable(s) Entered on Step Number
   1..    VAR00002

Multiple R            .70065
R Square              .49091
Adjusted R Square     .49020
Standard Error        .68408

Analysis of Variance
                     DF      Sum of Squares      Mean Square
Regression            1           324.00000        324.00000
Residual            718           336.00000           .46797

F = 692.35714          Signif F = .0000
```

```
------------- Variables in the Equation -------------
Variable          B           SE B        Beta          T       Sig T
VAR00002      .900000     .034204     .700649      26.313      .0000
(Constant)    .100000     .094707                   1.056      .2914

------------ Variables not in the Equation ------------
Variable      Beta In     Partial     Min Toler        T       Sig T
VAR00003      .164122     .218218      .900000      5.987      .0000
VAR00004      .051267     .071429      .988235      1.918      .0556

* * * * MULTIPLE REGRESSION * * * *
Equation Number 1    Dependent Variable..    VAR00001

Variable(s) Entered on Step Number
   2.. VAR00003

Multiple R            .71774
R Square              .51515
Adjusted R Square     .51380
Standard Error        .66806

Analysis of Variance
                   DF     Sum of Squares     Mean Square
Regression          2        340.00000       170.00000
Residual          717        320.00000          .44630

F = 380.90625           Signif F = .0000

------------- Variables in the Equation -------------
Variable          B           SE B        Beta          T       Sig T
VAR00002      .833333     .035210     .648749      23.668      .0000
VAR00003      .166667     .027836     .164122       5.987      .0000
(Constant)   -.166667     .102653                  -1.624      .1049

------------ Variables not in the Equation ------------
Variable      Beta In     Partial     Min Toler        T       Sig T
VAR00004      .000000     .000000      .803571       .000     1.0000

End Block Number 1    PIN = .050    Limits reached.
```

28.3 Interpreting the output in Table 28.2

- There is a great deal of information in Table 28.2. Multiple regression is a complex area and needs further study in order to understand all of its ramifications. In interpreting the results of this analysis we shall restrict ourselves to commenting on the following statistics: Multiple R, R Square,

Adjusted R Square and Beta. Most of these are dealt with in a simple fashion in *ISP* Chapter 28.

- **The predictor that is entered on the first step of the stepwise analysis is the predictor which has the highest correlation with the criterion. In this example this predictor is VAR00002.**
- **As there is only one predictor in the regression equation on the first step, Multiple R is a single correlation coefficient. In this case it is 0.70065 or 0.70 to two decimal places.**
- **R Square is the multiple correlation coefficient squared which in this instance is 0.49091 or 0.49 to two decimal places. This indicates that 49% of the variance in the criterion is shared with or 'explained by' the first predictor.**
- **Adjusted R Square is R Square which has been adjusted for the size of the sample and the number of predictors in the equation. The effect of this adjustment is to reduce the size of R Square so Adjusted R Square is 0.49020 or 0.49 to two decimal places. In this example R Square and Adjusted R Square are the same when rounded to two decimal places.**
- Beta in the Variables in the Equation section is the standardised regression coefficient which is the same as the correlation coefficient when there is only one predictor. It is as if all your scores had been transformed to *z*-scores before the analysis began.
- **The variable which is entered second in the regression equation is the predictor which generally explains the second greatest significant proportion of the variance in the criterion. In this example, this variable is VAR00003.**
- **The Multiple R, R Square and Adjusted R Square are 0.71774, 0.51515 and 0.51380 respectively which rounded to two decimal places are 0.72, 0.52 and 0.51.**
- **In other words, the two variables VAR00002 and VAR00003 together explain or account for 51% of the variance in the criterion.**
- Since we know how much of the criterion variance the first predictor (VAR00002) explains (49%), we can work out how much of the criterion variance the second predictor (VAR00003) explains by subtracting the first Adjusted R Square from the second Adjusted R Square. When we do this, we see that the second predictor (VAR00003) explains a further 2% (0.51 – 0.49 = 0.02).
- Beta is 0.648749 for the first predictor (VAR00002) and 0.164122 for the second predictor (VAR00003).
- The analysis stops at this point as the third predictor (VAR00004) does not explain a further significant proportion of the criterion variance. Notice that in the final few lines of the output, VAR00004 has a T value of 0.000 and the significance of T (Sig T) is 1.0000. This tells us that VAR00004 is a non-significant predictor of the criterion (VAR00001).

Table 28.3 **Stepwise multiple regression of predictors of VAR00001 (only significant predictors are included)**

Variable	Multiple R	B	Standard error B	Beta	T	Significance of T
VAR00002	0.70	0.83	0.04	0.65	23.67	0.001
VAR00003	0.72	0.17	0.03	0.16	5.99	0.001

28.4 Reporting the output in Table 28.2

- There are various ways of reporting the results of a stepwise multiple regression analysis. In such a report we should include the following kind of statement: 'In the stepwise multiple regression, intellectual ability explained 49% of the variance in educational achievement and school motivation a further 2%. Greater educational attainment was associated with greater intellectual ability and school motivation.'.
- A table is sometimes presented. There is no standard way of doing this but Table 28.3 is probably as clear as most.

Chapter 29

Hierarchical multiple regression

Hierarchical multiple regression allows the researcher to decide which order to use for a list of predictors. Rather than let the computer decide on the basis of statistical criteria, the researcher decides which should be the first predictor, the second predictor, and so forth. This order is likely to be chosen on theoretical grounds.

29.1 Introduction

We will illustrate the computation of a hierarchical multiple regression analysis with the data shown in Table 29.1 which consist of scores for six individuals on the four variables of educational achievement, intellectual ability, school motivation and parental interest respectively.

We have added a further variable, social class, which is on a scale of 1 to 5, with 5 being the highest social class. Hierarchical analysis is used when variables are entered in an order predetermined by the researcher on a 'theoretical' basis rather than in terms of statistical criteria. This is done by ordering the independent variables in terms of blocks of the independent variables, called block 1, block 2, etc. A block may consist of just one independent variable or several. In this particular analysis, we will make Block 1 social class (VAR00005) which is essentially a demographic variable which we would like to control for. Block 2 is going to be intellectual ability (VAR00002). Block 3 is going to be school

Table 29.1 **Data for hierarchical multiple regression**

Educational achievement	Intellectual ability	School motivation	Parental interest	Social class
1	2	1	2	2
2	2	3	1	1
2	2	3	3	5
3	4	3	2	4
3	3	4	3	3
4	3	2	2	2

motivation (VAR00003) and parental interest (VAR00004). The dependent variable or criterion to be explained is educational achievement (VAR00001).

In our example, the model essentially is that educational achievement is affected by intellectual ability which is partly determined by motivational factors such as school motivation and parental interest. Social class is being controlled for in this model since we are not regarding it as a psychological factor.

When doing a path analysis, it is necessary to do several hierarchical multiple regressions. One re-does the hierarchical multiple regression using different blocks and in different orders so that various models of the interrelationships can be explored.

29.2 Hierarchical multiple regression analysis

Quick summary

Enter data

Statistics

Regression

Linear . . .

Criterion and ▶ *beside Dependent:*

Predictors and ▶ *beside Independent[s]:*

OK

	var00001	var00002	var00003	var00004	var00005	var00006	var
1	1.00	2.00	1.00	2.00	2.00	20.00	
2	2.00	2.00	3.00	1.00	1.00	20.00	
3	2.00	2.00	3.00	3.00	5.00	20.00	
4	3.00	4.00	3.00	2.00	4.00	20.00	
5	3.00	3.00	4.00	3.00	3.00	20.00	
6	4.00	3.00	2.00	2.00	2.00	20.00	
7							

Figure 29.1 Five variables and weighting factor in Newdata

- Enter the data used in Chapter 28 in Newdata or retrieve the file if the data have been stored.
- Add the data for social class which is shown as a fifth variable in Figure 29.1. (The six rows have been weighted or re-entered several times to produce a sizeable number of cases. *You would not do this normally. It is merely a way of making the burden easier* and should only be used when reproducing our example.)
- Select Statistics from the menu bar in the Applications window which produces a drop-down menu (Figure 2.6).
- Select Regression from the drop-down menu which reveals a smaller drop-down menu.
- Select Linear . . . from this drop-down menu which opens the Linear Regression dialog box (Figure 8.1).
- Select 'var00001' and then the ▶ button beside the Dependent: box which puts 'var00001' in this box.
- Select 'var00005' and then the ▶ button beside the Independent[s]: box which puts 'var00005' in this box.
- Select Next. This makes Block 1 which consists only of 'var00005'. Keep an eye on the way in which the rectangle reading Block 1 of 1 changes over the next few steps. This will help you understand how the blocks are formed.
- Select 'var00002' and then the ▶ button beside the Independent[s]: box which puts 'var00002' in this box.
- Select Next. This makes Block 2 which consists of only 'var00002'.
- Select 'var00003' and 'var00004' and then the ▶ button beside the Independent[s]: box which puts these two variables in this box. We do not need to select Next this time, but essentially we have made Block 3 which consists of the two variables, 'var00003' and 'var00004'.
- Select OK which closes the Linear Regression dialog box and the Newdata window and which displays the output shown in Table 29.2 in the Output window.

29.3 Interpreting the output in Table 29.2

- The variable entered on the first block is VAR00005 (social class). The Adjusted R Square for this block is effectively 0.0 (–0.00424) which means that social class explains 0% of the variance of educational achievement.
- The statistical significance of the *F*-ratio (Signif F) for this block is 0.4821. As this value is above the required value of 0.05, this means that the regression equation at this first stage does not explain a significant proportion of the variance in educational achievement. Notice that the Regression sum of

Table 29.2 **Hierarchical multiple regression analysis output**

```
                  * * * * MULTIPLE REGRESSION * * * *

Listwise Deletion of Missing Data

Equation Number 1    Dependent Variable..    VAR00001

Block Number 1. Method: Enter    VAR00005

Variable(s) Entered on Step Number
   1..    VAR00005

Multiple R            .06478
R Square              .00420
Adjusted R Square    -.00424
Standard Error        .96348

Analysis of Variance
                    DF      Sum of Squares      Mean Square
Regression           1              .46154           .46154
Residual           118           109.53846           .92829

F = .49719          Signif F = .4821

------------------ Variables in the Equation ------------------

Variable              B          SE B        Beta          T      Sig T
VAR00005        .046154       .065456     .064775       .705      .4821
(Constant)     2.369231       .205256                 11.543      .0000

---------------- Variables not in the Equation ----------------

Variable        Beta In       Partial   Min Toler          T      Sig T
VAR00002        .739574       .712055     .923077     10.970      .0000
VAR00003        .395188       .370850     .876923      4.319      .0000
VAR00004        .179435       .118511     .434389      1.291      .1993

End Block Number 1    All requested variables entered.

                  * * * * MULTIPLE REGRESSION * * * *

Equation Number 1    Dependent Variable..    VAR00001

Block Number 2. Method: Enter    VAR00002

Variable(s) Entered on Step Number
   2..    VAR00002

Multiple R            .71351
R Square              .50909
Adjusted R Square     .50070
Standard Error        .67937
```

Analysis of Variance

	DF	Sum of Squares	Mean Square
Regression	2	56.00000	28.00000
Residual	117	54.00000	.46154

F = 60.66667 Signif F = .0000

------------- Variables in the Equation -------------

Variable	B	SE B	Beta	T	Sig T
VAR00005	−.100000	.048038	−.140346	−2.082	.0396
VAR00002	.950000	.086603	.739574	10.970	.0000
(Constant)	.250000	.241390		1.036	.3025

------------- Variables not in the Equation -------------

Variable	Beta In	Partial	Min Toler	T	Sig T
VAR00003	.223803	.290129	.825000	3.265	.0014
VAR00004	.358870	.333333	.395604	3.808	.0002

End Block Number 2 All requested variables entered.

* * * * MULTIPLE REGRESSION * * * *

Equation Number 1 Dependent Variable.. VAR00001

Block Number 3. Method: Enter VAR00003 VAR00004

Variable(s) Entered on Step Number
3.. VAR00003
4.. VAR00004

Multiple R	.76871
R Square	.59091
Adjusted R Square	.57668
Standard Error	.62554

Analysis of Variance

	DF	Sum of Squares	Mean Square
Regression	4	65.00000	16.25000
Residual	115	45.00000	.39130

F = 41.52778 Signif F = .0000

------------- Variables in the Equation -------------

Variable	B	SE B	Beta	T	Sig T
VAR00005	−.312500	.067717	−.438581	−4.615	.0000
VAR00002	.937500	.083853	.729843	11.180	.0000
VAR00003	.187500	.067717	.184637	2.769	.0066
VAR00004	.437500	.129668	.314011	3.374	.0010
(Constant)	−.562500	.283550		−1.984	.0497

End Block Number 3 All requested variables entered.

squares + the Residual sum of squares add to 110. This total sum of squares is the same for all of the succeeding Analyses of Variance for later steps in the regression equation.

- The variable entered on the second block is VAR00002 (intellectual ability). The Adjusted R Square for this block is 0.50070 which means that intellectual ability together with social class explain 50.07% of the variance of educational achievement.
- The statistical significance of the *F*-ratio for this block is 0.0000. As this value is much lower than the critical value of 0.05, the first two steps of the regression equation explain a significant proportion of the variance in educational achievement. Notice that the sums of squares for the Regression and the Residual again add to 110.
- The variables entered on the third and final block are VAR00003 (school motivation) and VAR00004 (parental interest). The Adjusted R Square for this block is 0.57668 which means that all four variables explain 57.69% of the variance of educational achievement.
- The statistical significance of the *F*-ratio for this block is 0.0000. As this value is much lower than the critical value of 0.05, the first three steps in the regression equation explain a significant proportion of the variance in educational achievement.
- **The simplest interpretation of Table 29.2 comes from examining the final table, headed Variables in the Equation. Especially useful are the Beta column and the Sig T column. These tell us that the correlation (Beta weight) between VAR00005 (social class) and VAR00001 (educational achievement) is –0.438581 which is significant at the 0.0000 level. Having controlled for social class, Block 1, the correlation between VAR00002 (intellectual ability)**

Table 29.3 **Hierarchical multiple regression of predictors of educational achievement (only significant predictors are included)**

Blocks	B	Standard error B	Beta
Block 1:			
Social class	–0.31	0.07	–0.44*
Block 2:			
Intellectual ability	0.94	0.08	0.73*
Block 3:			
School motivation	0.19	0.07	0.18*
Parental interest	0.44	0.13	0.31*

* Significant at 0.001.

and VAR00001 (educational achievement) is 0.729843. This is also significant at the 0.0000 level. Finally, having controlled for VAR00005 (social class) andVAR00002 (intellectual ability), the correlations for each of the variables in Block 3 (school motivation and parental interest) with educational achievement (VAR00001) are given separately.

29.4 Reporting the output in Table 29.2

There are various ways of reporting the results of a hierarchical multiple regression analysis. In such a report we would normally describe the percentage of variance explained by each set or block of predictors. One would also need to summarise the regression equation as in Table 29.3.

Appendix

Other statistics on SPSS

Other statistical tests provided by SPSS but not described in this book are shown below in terms of their options on the Statistics menu, submenu and dialog box options.

Statistics menu	Statistics submenu	Dialog box
Summarize	Crosstabs . . .	Lambda Uncertainty coefficient Gamma Somers' d Kendall's tau-b Kendall's tau-c Kappa Risk Eta
Correlate	Bivariate . . .	Kendall's tau-b
Regression	Curve Estimation . . . Logistic . . . Probit . . . Nonlinear . . . Weight Estimation . . . 2-Stage Least Squares . . .	
Loglinear	General . . . Logit . . . Hierarchical . . . (Model Selection . . .)	
Classify	K-Means Cluster . . . Hierarchical Cluster . . . Discriminant . . .	

Scale	Reliability Analysis . . .	
	Multidimensional Scaling . . .	
Nonparametric Tests	Runs . . .	
	1-Sample K-S . . . (Kolmogorov-Smirnov)	
	2 Independent Samples . . .	Kolmogorov-Smirnov Z
		Wald-Wolfowitz runs
		Moses extreme reactions
	K Independent Samples . . .	Kruskal-Wallis H
		Median
	K Related Samples . . .	Friedman
		Kendall's W
		Cochran's Q
Time Series	Exponential Smoothing . . .	
	Autoregression . . .	
	ARIMA . . .	
	X11 ARIMA . . .	
	Seasonal Decomposition . . .	
Survival	Life Tables . . .	
	Kaplan-Meier . . .	
	Cox Regression . . .	
	Cox w/ Time-Dep Cov . . .	

Index